Bruno Lucci
Wendell Lamas
Francisco J. Grandinetti

Analysis and Optimization of a Metal x Metal Seal

Bruno Lucci
Wendell Lamas
Francisco J. Grandinetti

Analysis and Optimization of a Metal x Metal Seal

using the Finite Element Method

ScienciaScripts

Imprint

Any brand names and product names mentioned in this book are subject to trademark, brand or patent protection and are trademarks or registered trademarks of their respective holders. The use of brand names, product names, common names, trade names, product descriptions etc. even without a particular marking in this work is in no way to be construed to mean that such names may be regarded as unrestricted in respect of trademark and brand protection legislation and could thus be used by anyone.

Cover image: www.ingimage.com

This book is a translation from the original published under ISBN 978-613-9-65614-1.

Publisher:
Sciencia Scripts
is a trademark of
Dodo Books Indian Ocean Ltd. and OmniScriptum S.R.L publishing group

120 High Road, East Finchley, London, N2 9ED, United Kingdom
Str. Armeneasca 28/1, office 1, Chisinau MD-2012, Republic of Moldova, Europe
Printed at: see last page
ISBN: 978-620-7-79279-5

SUMMARY

I dedicate this work to my dear wife Léia, and to my parents, Luiz Francisco and Anette.

To my advisor Wendell, for his commitment to guiding me.

ACKNOWLEDGMENTS

To the University of Taubaté, for the opportunity to continue my studies in the field of engineering.

To the professors of the Master's in Mechanical Engineering, for the knowledge shared with the students.

To my colleagues in the Professional Master's in Mechanical Engineering class at the University of Taubaté, for their friendship and for sharing information and experiences.

To Prof. Dr. Wendell de Queiróz Lamas, for the skill with which he guided our work.

"The essence of knowledge consists in applying it once possessed."
Confucius

SUMMARY

Subsea equipment used for oil and gas exploration uses metal-to-metal seals to seal its components from the environment. Historically, the design of these seals has been developed using previous experience of seals already in use, without carrying out a calculation analysis on these components. After the design of the seals has been developed and before they are used, their efficiency must be proven through validation tests. This work presents and suggests the use of finite element *software* as a tool for developing these seals, which aims to reduce the number of designs and empirical tests, increasing the probability of success during validation tests, reducing the number of real tests required, which has the effect of reducing execution costs. In developing this work, it was shown how to configure a seal model in the finite element *software*, different mesh sizes were simulated to assess which configuration presented results and computational costs compatible with the work developed in order to proceed with the development of the other models, several models were developed iteratively, that is, according to the results presented at the end of each model, changes were made to the geometry in order to optimize the results and at the end of each model the results were evaluated. In the end, the six models were developed and evaluated and the one chosen for use was the one with sufficient contact tension to perform sealing, tensions compatible with the failure criterion adopted, plastic deformation in the seal within an acceptable range, and behavior of the system as suggested by Sweeney et al. (2004), at the ends of the seal.

Keywords: Finite element model; "metal x metal" seals; subsea equipment.

CHAPTER 1 - INITIAL CONSIDERATIONS

1.1. INTRODUCTION

The equipment used to prospect for oil and gas uses metal-to-metal seals to ensure that its components are watertight against the external environment. According to international standards (American Petroleum Institute, 2011, 2010), metal-to-metal seals are qualified through specific validation tests to guarantee their performance under different working pressure conditions. Figure 1 shows a DX-type subsea connector as an example, where "metal x metal" seals are used to seal the system from the environment.

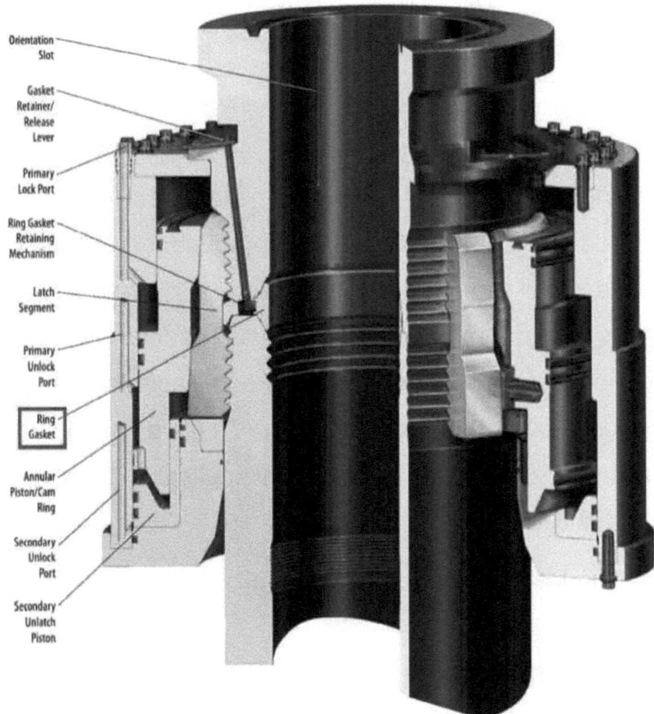

Figure 1 - Example of a DX-type subsea connector using a metal-to-metal seal - sectional view (Dril-Quip, 2015).

The DX subsea connector is a component that connects various types of equipment to the wellhead. The system's interfaces are sealed by metal-to-metal seals to guarantee a seal between the environment and the well.

For sealing to occur, the mechanical contact between two surfaces must prevent the flow of mass between them.

To achieve a seal, it is necessary to fill the free space between the two surfaces in contact so that there is no flow path. This mechanical impediment can be achieved by inserting a seal between the two surfaces to be sealed or even by compacting one surface against the other (Brandâo, 2007). The best known types of seals are elastomeric and metal x metal. The best known elastomeric seals are *o-rings* and gaskets. Among the best known metal-to-metal subsea seals are the BX type and specific seals developed by subsea equipment manufacturers. Metal x metal seals are preferably used in subsea equipment as primary seals, due to their greater durability and reliability.

Historically, validation tests for metal x metal seals were carried out using the trial and error method, where the pre-design of the seal was based on previous experience and the experience of the designer and then validation tests were carried out. When the validation tests didn't produce the expected results, the designer made small changes to the design, a new seal was manufactured and the tests were repeated. This cycle was repeated until the seal passed all the validation tests. The validation tests applicable to metal x metal subsea seals preferably follow those described in API 6A and API 17D (American Petroleum Institute, 2011, 2010). The main applicable tests are pressure and temperature tests, as shown in Table 1.

Table 1 - Minimum validation test requirements for metal x metal seals used in wellhead or spindle systems
seals used in wellhead or Christmas tree systems (American Petroleum Institute, 2011, p. 3, Table 3).
(American Petroleum Institute, 2011, p. 24, Table 3).

Component	Cyclic pressure /	Cyclic temperature test	Cyclic endurance test

	load test		
Metal seal exposed to the fluid produced by the well	200	3	As defined by the manufacturer
Metal seal not exposed to the fluid produced by the well	3	3	As defined by the manufacturer

1.2. THEORETICAL BASIS

1.3. 1.The finite element method

Azevedo (2003) will show the basic concepts of the finite element method, so that the user of this method understands how it works and can achieve reliable results as a result of using it.

The work of Azevedo (2003) will help in understanding the application of the finite element method. In this work in particular, it helps to understand the correlation between the assembly of a finite element system and the model mesh.

Alves Filho (2007) argues that in order to use the finite element method and related *software* correctly, the professionals involved must have a fundamental conceptual basis. Using a graphical tool such as specialized finite element *software* without the correct conceptual basis can lead to the quickest way to get wrong answers.

The work by Alves Filho (2007) explains in simple terms how the finite element method works, exemplifying how a part subdivided into elements can result in an understanding of the problem as a whole, since these elements are interconnected.

In his thesis, Augusto (2012) presents an object-oriented architecture for the high-order finite element method. Its purpose is to develop and explain the development of a finite element program for applications in structural mechanics.

Augusto's work (2012) will help you understand the theory of Finite Elements. It also provides a great explanation of the use of contact in FEM, explaining the different methods of configuration and convergence.

1.1.1.1. Using the finite element method to develop seals

Flach (1995) describes the stages that need to be followed in order to develop a commercially viable label. The summary of these steps is: identification and evaluation; feasibility; prototype development and laboratory testing; field testing; product launch and; monitoring the use of the product.

In the prototype development stage described by Flach (1995), the possible tools to be used are described, including computer modeling and the use of the finite element method. These two tools used together are capable of predicting the results of the actual use of the seals.

Anderson (1996) described how finite element analysis has revolutionized the design process for elastomeric seals, minimizing the cost of developing prototypes by trial and error, using the finite element method to predict and determine results before a prototype is actually manufactured.

This author's idea (Anderson, 1996) is in line with the main focus of this work, to suggest the use of the finite element method as a tool for developing seals, working on the development and optimization of the seal design before subjecting it to real use tests.

Hou et al. (2010) described in their article a finite element analysis of an *o-ring* seal in its *dovetail* housing. The results of the maximum pressure of the seal under different pressure and temperature conditions are shown graphically and in tabular form.

This article (Hou et al., 2010) contributes to this work in the sense of how to evaluate the stress results found graphically and interpret them correctly in the behavior of the seal during use. Even though it is an elastomeric seal, this article demonstrates that a seal should result in high stresses in the effective sealing regions.

Finney (2012) explained the use of the finite element method in systems using metal and rubber. In the introduction, he emphasizes that the use of the finite element method is only a tool and must be used correctly in order to obtain reliable results. After this introduction, the steps on how to set up the finite element model are described, such as configuring the materials, selecting the type of analysis, building the model, boundary conditions, solutions, results and analyzing the results.

In this work, the theory described by Finney (2012) helps us to understand how to correctly configure the model in order to obtain reliable results. It also reinforces the use of a non-linear stress x strain curve, as the stresses in this work tended to exceed the yield stress of the materials involved.

1.1.2. Seal development

In their patent, Sweeney et al. (2004) described a metal x metal seal used in a subsea wellhead system. As in any patent, they described the main characteristics of the product, such as the upper and lower legs of the seal, the unique sealing surface, among others.

This paper develops a seal with similar characteristics to the concept presented by Sweeney et al. (2004). This patent also reinforced the idea of using a conical seal, with a small angular difference between seal and seat, in order to increase the tension at the end of the seal during use, creating a sealing strip responsible for sealing the system.

The *American Petroleum Institute* is the world's largest petroleum organization. This institution has a committee of industry professionals who create and revise the most varied standards in this area, with the aim of standardizing the rules used in the sector, resulting in increased and guaranteed safety in projects and operations, environmental protection, respect for current legislation and cost efficiency in operations.

The API 6A and API 17D standards (American Petroleum Institute, 2011, 2010) were created to standardize the design, manufacturing, testing and quality requirements for wellhead systems and Christmas trees.

The API 17D standard defines which items must undergo validation testing, what the validation requirements are and which steps must be followed when carrying out the validation tests. This work was based on the requirement for validation tests for metal seals defined in this standard.

1.1.3. Failure criterion - von Mises stress

The material developed by Natal Jorge and Dinis (2005) described the theory of plasticity, aimed at the design of structural components. It presented the fundamental concepts of the elasto-plastic model: hardening, plastic flow and its rules.

The main contribution of this publication (Natal Jorge and Dinis, 2005) to this work is in the graphical understanding of the von Mises stress. It shows the region covered by the von Mises stress and a comparison with the Tresca stress, clearly explaining that when working with the material within the delimited area of the yield surface, the material is working in its elastic regime and when working in regions beyond this surface, the material is working in its plastic regime.

The article by Jong and Springer (2009) provides a comparative explanation of the usual simplified formulation used for the von Mises stress and the complete formulation mentioned in the most recent publications.

This publication (Jong and Springer, 2009) contributes to this work by providing a deeper understanding of the von Mises stress formulation. It shows that for basic studies the conventional formulation can be applied, but for more advanced and complete studies, it is best to use the full von Mises stress formulation.

Christensen (2015) has provided material on the main failure criteria in various types of materials, explaining that failure criteria depend on the type of material, type of material behavior, applicable loads, temperature conditions, among other variables.

The material provided by Christensen (2015), which compares the von Mises and Tresca failure criteria, helps to justify the choice and use of the von Mises failure criterion in this work.

Lima (2015) has made available material which explains the strength, yield/plasticity and rupture criteria of materials used in mechanical design. In this material, the different failure criteria are explained, and which ones are suggested to be considered for materials with different behaviors.

The work of Lima (2015) helps in this work in understanding the different failure criteria and in choosing the criterion used.

1.1.4. Mechanical contact and FEM contact

Johnson (1987) published in his book a continuation of the study of mechanical contacts. He uses the theory initially published by other authors, such as Hertz, and from there develops complex formulations on the most varied types of contact between materials, which are accepted and used by many researchers today.

Johnson's (1987) publication helps us to understand how contact stress works and which basic variables influence contact stress.

In his thesis, Bryant (2013) studied contact and deformation on the rough surfaces of gear teeth in detail. In this situation, he made a comparison between the application of the finite element method and the application of the formulas and theories found in Johnson's (1987) book. The results found in these comparisons were very close.

The development and citations of contact pressure theory found in this thesis (Bryant, 2013) help us to understand the variables that exist in the formulations that govern contact and what influence each of these has between two bodies in contact.

1.1.5. Project development costs

Bay and Bay (2010) cover general knowledge about subsea engineering, specifically for the oil and gas sector. Their work is of great value to professionals working in this area, covering a wide range of functions, such as engineering, project management and cost planning.

The book published by Bay and Bay (2010) has a specific chapter on cost estimation in subsea projects. In this work, a cost analysis was carried out, comparing the development of a seal with the aid of the finite element method and without the aid of the finite element method. This publication provided information that helped to gather data for the cost comparison between the methods.

1.3. OBJECTIVES

The purpose of the study is to verify the feasibility of using the finite element method to simulate qualification tests for "metal x metal" seals for use underwater in oil prospecting equipment.

This work suggests the use of specialized finite element calculation *software to* help with the design stages, simulating validation tests due to the high costs of manufacturing, testing and redesign. It also aims to define and show the steps required to carry out calculations using FEM for "metal x metal" seals.

Once the ideal theoretical design has been verified, the validation test would be carried out, increasing the likelihood of successful test results. Using the proposed methodology, real preliminary tests can be replaced by computer simulations, reducing time and cost.

1.4. BACKGROUND

This type of work is little publicized in literature and academic papers.

Therefore, this work aims to develop and show all the necessary steps for executing and calculating metal x metal seals using the FEM. It also aims to demonstrate that the use of FEM in the development of seals can bring economic benefits to this process.

1.5. MATERIALS AND METHODS

1.5.1. Methodology

This is exploratory research, as its purpose is to clarify issues relating to "metal x metal" seals in their actual conditions of use and suitability for the performance required by international standards (American Petroleum Institute, 2011, 2010) in qualification tests, with a quantitative approach, where the data considered and collected is treated using finite element analysis methods, which are inherently quantitative.

1.5.2. Modeling using the finite element method

The finite element method was used to develop this work. There are several finite element *software programs on* the market, and ABAQUS® was chosen for this work.

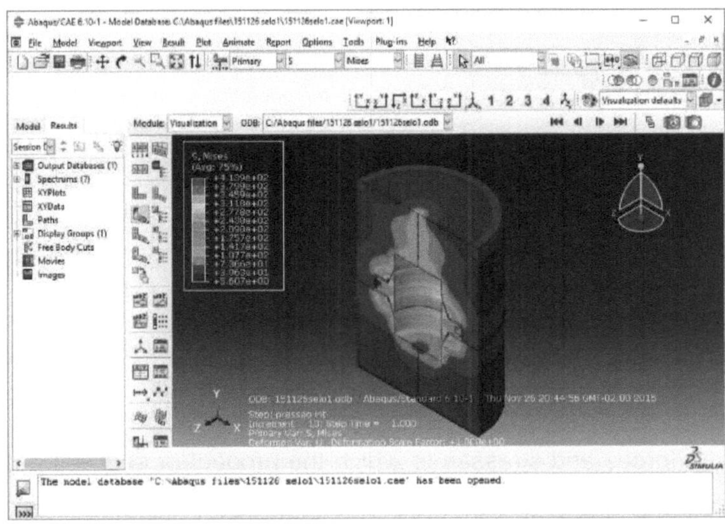

Figure 2 - Using the ABAQUS® *software.*

The Finite Element Method is an approximate method for calculating continuous systems in such a way that the structure, the mechanical component, or, in general, the continuous body, is subdivided into a finite number of parts (the elements), connected to each other through discrete points, which are called nodes. The assembly of elements, which constitutes the mathematical model, has its behavior specified by a finite number of parameters [...] (Alves Filho, 2007, p. 11).

These parameters, formed by differential equations, are solved until the expected results are obtained.

The FEM formulation requires the existence of an integral equation, so that it is possible to replace the integral over a complex domain (of volume V) by a sum of integrals extended to subdomains of simple geometry (of volume Vi). This technique is illustrated with the following example, which corresponds to the volume integral of a function f (Azevedo, 2003, p. 4).

The formula cited by Azevedo (2003) is illustrated by Equation (1).

$$\int_{v} f dV = \sum_{i=1}^{n} \int_{V_i} f dV \tag{1}$$

Equation (1) assumes that:

$$V = \sum_{i=1}^{n} V_i \tag{2}$$

If it is possible to calculate all the integrals extended to the subdomains Vi, it is sufficient to perform the summation corresponding to the second member of Equation (1) to obtain the integral extended to the entire domain. Each subdomain Vi corresponds to a finite element of simple geometry [...] The sum indicated in Equation (1) will give rise to the operation known as assembly (Azevedo, 2003, p. 5).

There are various numerical methods for solving problems, but the finite element method is one of the most suitable methods for solving complex design problems. As geometries become more complex, the techniques and theories used in an attempt to simplify the problem and define an equation are obviously inapplicable, or impossible for the design engineer to solve. FEM allows the analysis of these complex structures without the need to develop and apply complex equations (Finney, 2012, p. 259).

Once a detailed model of the seal has been developed, a detailed understanding of the forces and stresses to which the product or seal will be subjected is required. Ranges of operating conditions can be calculated, such as fixed or variable pressure and temperature conditions or movement of components around the seal.

Manufacturers of mechanical seals have been developing state-of-the-art modeling programs that allow designers to make changes to products and analyze their effects quickly and efficiently (Flach, 1995).

Now a project can be conceived, designed and modeled by computer using finite element calculation before components are manufactured. This allows design interactions, results and even design optimization to be carried out before manufacturing begins. Although tests are still necessary to verify the results, greater predictability and components with lower production costs can be developed in a shorter time (Anderson, 1996).

Finite element analysis was used to determine the optimum geometry for the seal face and seat under a specific operating pressure, temperature and speed condition (EagleBurgmann Germany Gmbh & Co KG, 2012).

Accounting for a myriad of possibilities can be complex and must be done on a case-by-case basis, requiring human verification and experienced operators (Trelleborg Sealing Solutions (TSS), 2013).

1.5.3. The material failure criterion

In addition to checking the contact pressure, a basic output in seal calculation is to check the material for failure criteria.

If a tensile test is carried out on a specimen of a ductile material, it can be said that the specimen fails when the axial stress reaches the yield stress, i.e. the failure criterion is yielding (Lima, 2015).

According to the characteristics of the materials considered in this study, they are classified as ductile. For these types of materials, the failure criteria used are: the von Mises criterion or the Tresca criterion.

1.5.3.1. The von Mises criterion

This criterion is based on determining the distortion energy of a given material, i.e. the energy related to changes in the shape of the material [...]. According to this criterion, a component is safe as long as the highest value of distortion energy per unit volume of the material remains below the distortion energy per unit volume required to cause yielding in a specimen of the same material subjected to a tensile test (Beer et al., 2007, p. 637-638).

The von Mises stress (σ') is calculated from Equation (3).

$$\sigma' = \frac{1}{\sqrt{2}} \cdot \left[\left(\sigma_x - \sigma_y\right)^2 + \left(\sigma_y - \sigma_z\right)^2 + \left(\sigma_z - \sigma_x\right)^2 + 6 \cdot \left(\tau_{xy}^2 + \tau_{yz}^2 + \tau_{zx}^2\right) \right]^{\frac{1}{2}} \qquad (3)$$

where σ_x, σ_y, and σ_z are the normal stresses in the principal axes and τ_{xy}, τ_{yz}, and τ_{zx} are the shear stresses in the principal planes (Jong and Springer, 2009).

1.5.3.2. The Tresca criterion

The Tresca criterion is based on the maximum shear stress that the material can withstand until it fails.

The Tresca stress (σ') is calculated by Equation (4) for ordered principal stresses and by Equation (5) for unordered principal stresses.

$$\sigma' = \frac{1}{2}\left(\sigma_1 - \sigma_3\right) \qquad (4)$$

$$\sigma'^2 = \frac{1}{4}\left(\sigma_1 - \sigma_2\right)^{\frac{1}{2}}$$
$$\sigma'^2 = \frac{1}{4}\left(\sigma_2 - \sigma_3\right)^{\frac{1}{2}} \qquad (5)$$
$$\sigma'^2 = \frac{1}{4}\left(\sigma_3 - \sigma_1\right)^{\frac{1}{2}}$$

where σ_x, σ_y, and σ_z are the principal stresses (Christensen, 2015).

1.5.3.3 Comparison between failure criteria and the criterion adopted

The main difference between the criteria is that the von Mises stress represents the critical value of distortional energy stored in an isotropic material, while the Tresca stress represents the maximum value of shear stress in the isotropic material (Christensen, 2015). Figure 3 shows the comparison between the von Mises and Tresca stresses.

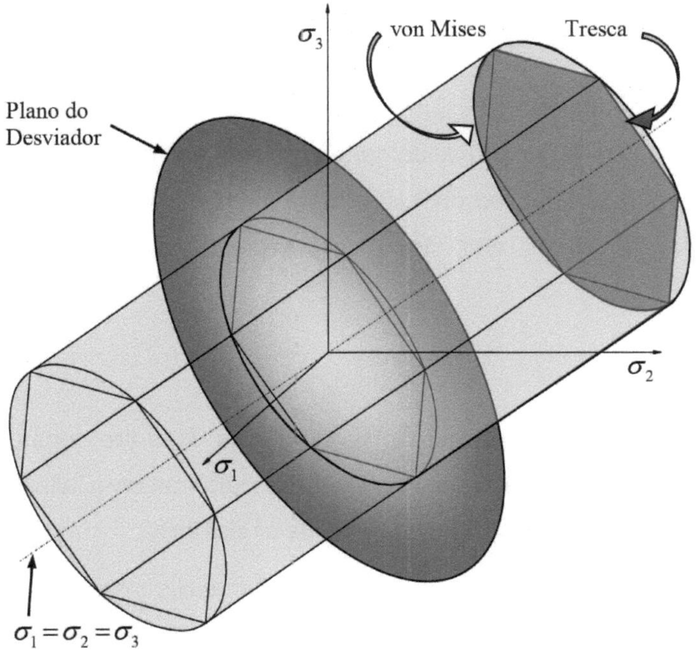

Figure 3 - Graphical representation of the flow surfaces for the Tresca and von
Mises criteria
(Natal Jorge and Dinis, 2005, p. 28).

Figure 3 shows the comparison between the Tresca and von Mises flow
criteria. The maximum allowable surface of each method can be seen. While
each flat section generated by the diverter plane generates a surface
delimited by a hexagon for the Tresca stress, in the same plane for the von
Mises stress the area is delimited by a regular elliptical figure, which touches
all the ends of the Tresca hexagon. According to each method, if a stress is
applied within the areas delimited by the surfaces, the material will work in its
elastic phase. However, if the stress exceeds the limits of these surfaces, the
material will flow and work in its plastic phase.

According to Beer et al. (2007, p. 638), the Tresca stress criterion is more
conservative. However, the von Mises criterion gives more accurate values
when used with ductile materials. In addition, the American Petroleum

17

Institute (2010) adopts the von Mises criterion. For these reasons, the von Mises criterion was considered in this work.

1.5.4 Permanent deformation

The degree to which a structure deforms or tensions depends on the magnitude of the stress imposed on it. For most metals that are stressed at relatively low levels, stress and strain vary proportionally as shown in Equation (6).

$$\sigma = E \cdot \varepsilon \tag{6}$$

This is known as Hooke's law, where σ is the stress, E is the constant of proportionality or modulus of elasticity and ε is the deformation (Callister Jr., 2006, p. 137). Each material has its own modulus of elasticity.

Figure 4 shows the typical stress-strain behavior of metallic materials. When a stress is applied with a value less than or equal to point P, the material works in its elastic phase and, after the stress is released, the material recovers its original condition. If the applied stress exceeds point P, the material will undergo permanent deformation and will not recover its original condition once the stress has been released. The yield stress of the material is represented by σ_y.

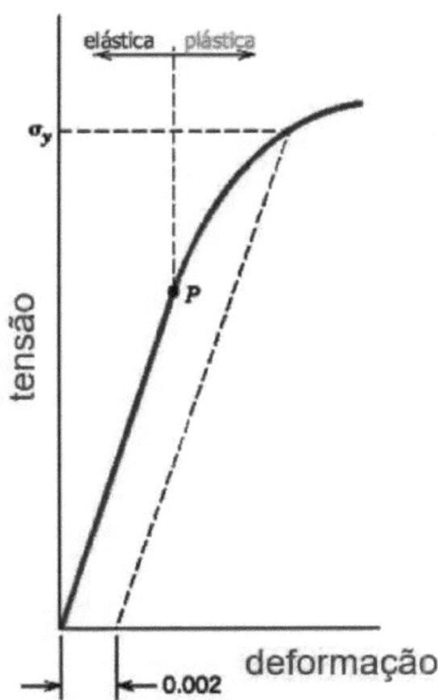

Figure 4 - Typical stress-strain behavior for metallic materials (Callister Jr., 2006, p. 143).

1.5.5. Contact pressure

1.5.5.1. Mechanical contact

The first satisfactory analysis of contact pressure between two elastic solids in contact was presented by Hertz (Johnson, 1987). Based on this theory, solutions for contact lines were developed by Johnson (1987). A summary of the formulation for calculating the key details for point and line contact is shown in Table 2. In this case, the contact point is understood to be the point of contact between a sphere and a plane; while the contact line is understood to be the line of contact between two cylinders or a cylinder and a plane.

Table 2 - Elastic contact formulation for point and line contact (Bryant, 2013, p. 70).

Description	Point of Contact	Contact Line

Size of contact	$a = \sqrt[3]{\dfrac{3}{2} \cdot \dfrac{R \cdot w}{E'}}$	$a = \sqrt{\dfrac{8}{\pi} \cdot \dfrac{R \cdot w'}{E'}}$
Maximum contact pressure, po	$p_0 = \dfrac{3 \cdot w}{2 \cdot \pi \cdot a^2}$	$p_0 = \dfrac{2 \cdot w'}{\pi \cdot a}$
Contact / interference approach distance, ω	$\omega = \dfrac{a^2}{R}$ $\omega \cong 1.31 \cdot \sqrt[3]{\left(\dfrac{w^2}{R \cdot E'^2}\right)}$	N/A
Where	$\dfrac{2}{E'} = \dfrac{1 - v_1^2}{E_1} + \dfrac{1 - v_2^2}{E_2}$	$\dfrac{1}{R} = \dfrac{1}{R_1} + \dfrac{1}{R_2}$

Where E_1 is the modulus of elasticity of the first body in contact, E_2 is the modulus of elasticity of the second body in contact, *E' is the* average modulus of elasticity calculated using the expression in Table 2, v_1 is the Poisson's ratio of the first body in contact, v_2 is the Poisson's ratio of the second body in contact, R_1 is the contact radius of the first body in contact, R_2 is the contact radius of the second body in contact, R is the average contact radius calculated from the expression in Table 2, *w is the* applied load and *w' is the* applied load per unit length.

1.5.5.2. Contact in FEM

When FEM is used on models with two or more parts in contact, it is often not possible to precisely define the area where contact occurs. This definition is even more difficult when using models with large deformations and displacements between parts, which is the case with the model developed in this work. Furthermore, contacts are often more complex than just the contact point or contact line.

The penalty method, mentioned in Augusto's work (2012), is recommended when the contact region is not known *a priori.* This method estimates the displacement that will occur in the increment of time and/or load applied in a given stage, and this process is iterative until the *software* achieves convergence of results. In our work, the *penalty* method was used for both normal and tangential contact.

CHAPTER 2 - DEVELOPMENT

2.1. CONTEXT

The development of a new seal begins when there is a demand to develop subsea equipment and new pressure conditions, different passage sizes, and/or new material requirements are required. In this specific case, the development of a new seal with the conditions set out in Table 3 is required.

Table 3 - Boundary conditions.

Description of the condition	Value
Maximum internal working pressure	69 MPa
Minimal internal circular passage	130.17 mm (5.125 in)

New developments for subsea oil and gas equipment generally require pressures of 69 MPa or more. The minimum circular internal passage in the order of 130 mm is a common value used for wet Christmas trees and related equipment on production lines.

2.2. MODEL OPTIMIZATION

For this work, an interactive development method was used. Six models were developed. After completing each model, the results were checked and the applicability of modifying geometric parameters to perform the calculation of a new model was evaluated. The parameters chosen were the differential angle and the geometric interference of the seal. These parameters are explained in more detail below.

2.3. MODEL GEOMETRY

The main components of a metal-to-metal sealing system are the seals or sealing rings and seats (Figure 5).

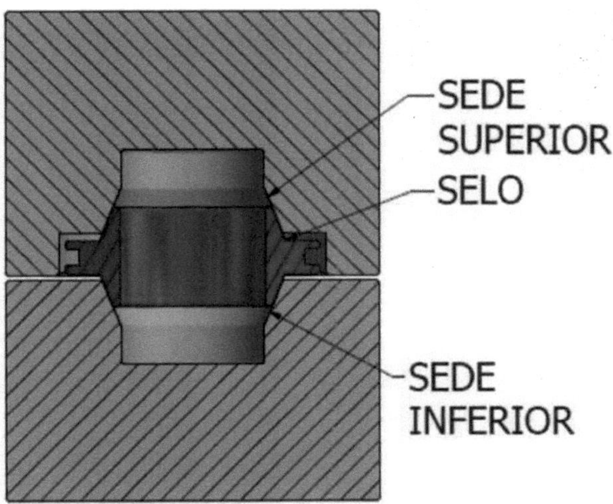

Figure 5 - Main components of the "metal x metal" sealing system.

An axisymmetric finite element model was created for each proposed geometry using the ABAQUS® *software* distributed by Dassault Systemès (2015).

The main dimensions of the components are shown in Figures 6 and 7. The geometry of the first model generated in the ABAQUS® *software* is shown in Figure 8.

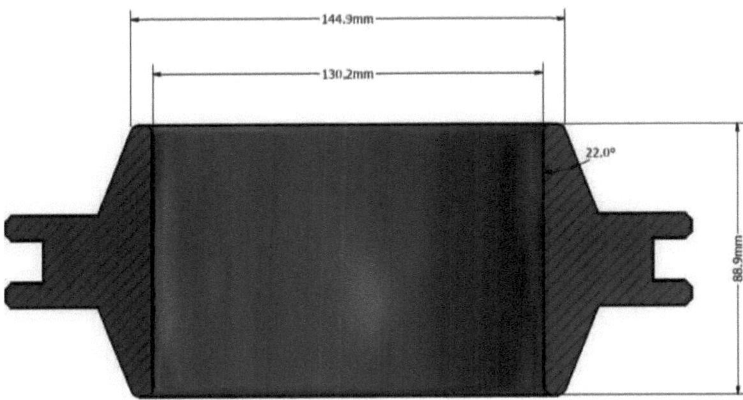

Figure 6 - Main dimensions of the seal - first model.

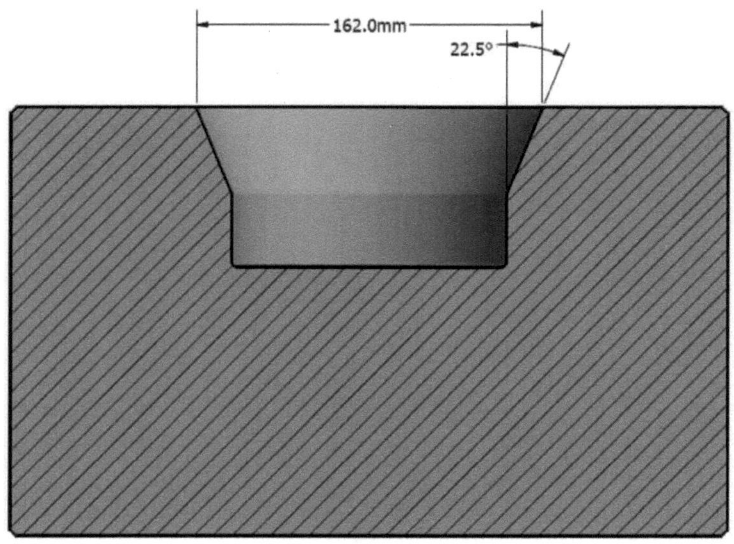

Figure 7 - Main dimensions of the headquarters - first model.

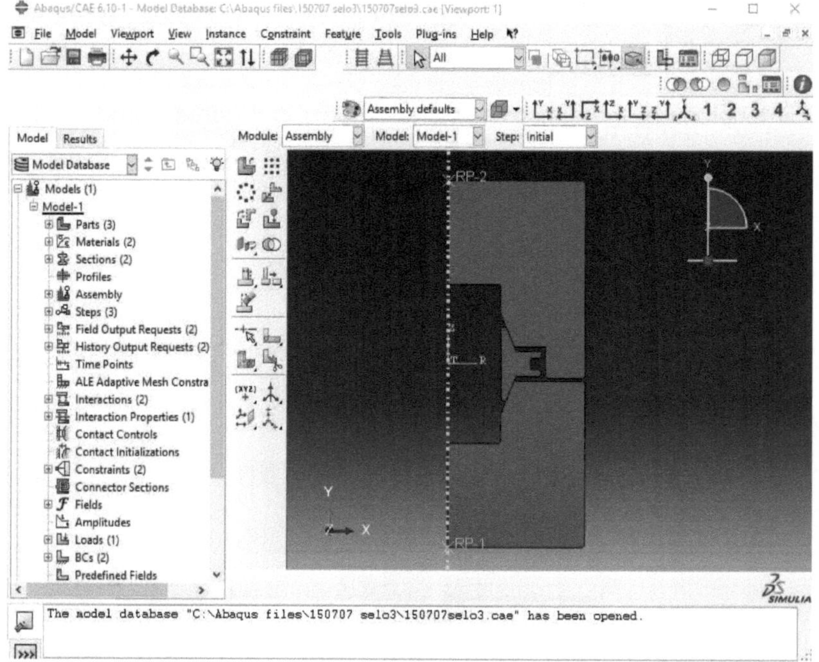

Figure 8 - Axisymmetric geometry of the first model created using ABAQUS® *software*.

1.1.1. Energizing course

The energizing stroke is the relative displacement made between the seats in order to deform the seal. The expected result of this energization process is that the system presents a contact tension on the contact surfaces sufficient to seal the system. Two energization stroke values were used as a basis: 3.5 mm and 2.3 mm. However, due to geometry adjustments, small variations occurred. The values used are shown in Table 4. The energizing stroke is shown in Figure 9.

Table 4 - Power stroke and differential angle used in each model.

Model	Energizing stroke (mm) Ân(	differential value (°)
First Model	3,64	0,5
Second Model	3,44	1,5
Third Model	2,31	1
Model Room	3,54	1
Fifth Model	2,21	1,5
Sixth model	2,41	0,5

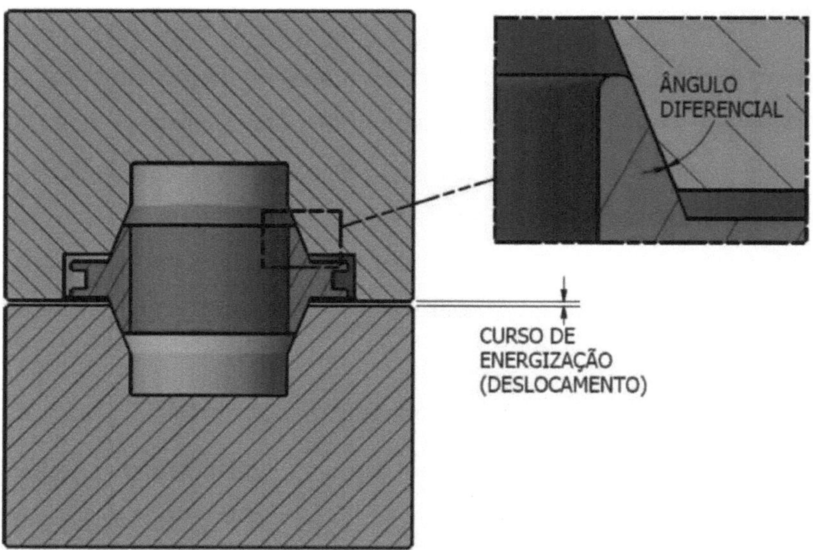

Figure 9 - Power stroke and differential angle of the model.

1.1.2. Differential Angle

The differential angle or initial relief angle was chosen empirically, following the recommendations of Sweeney et al. (2004) where it is described that the conical angle between the seal and the seat is preferably slightly different, in order to create a greater contact tension in the sealing strip. The values used are shown in Table 4. The differential angle is shown in Figure 9.

2.4. MATERIAL PROPERTIES

The materials chosen for the headquarters and the seal for this work are described in Table 5.

Table 5 - Materials chosen.

Component	Material
Sealing ring (seal)	Alloy 625
Headquarters	League 825

These materials were chosen to be compatible with the internal working fluids. Both materials have good corrosion resistance against aggressive environments, such as the fluids produced in subsea oil and gas systems.

The material properties for alloys 625 and 825 are shown in Tables 6 and 7. The input of the material properties into the *software* is shown in Figures 10 and 11.

Table 6 - Properties of alloy 625 (Special Metals Corporation, 2006).

Properties	Value
Modulus of elasticity	207.5 GPa
Poisson's ratio	0,278
Flow tension	413.68 MPa
Tensile strength	827.37 MPa

Table 7 - Properties of alloy 825 (Special Metals Corporation, 2004).

Properties	Value
Modulus of elasticity	196.0 GPa
Poisson's ratio	0,29
Yield stress	324 MPa
Tensile strength	690 MPa

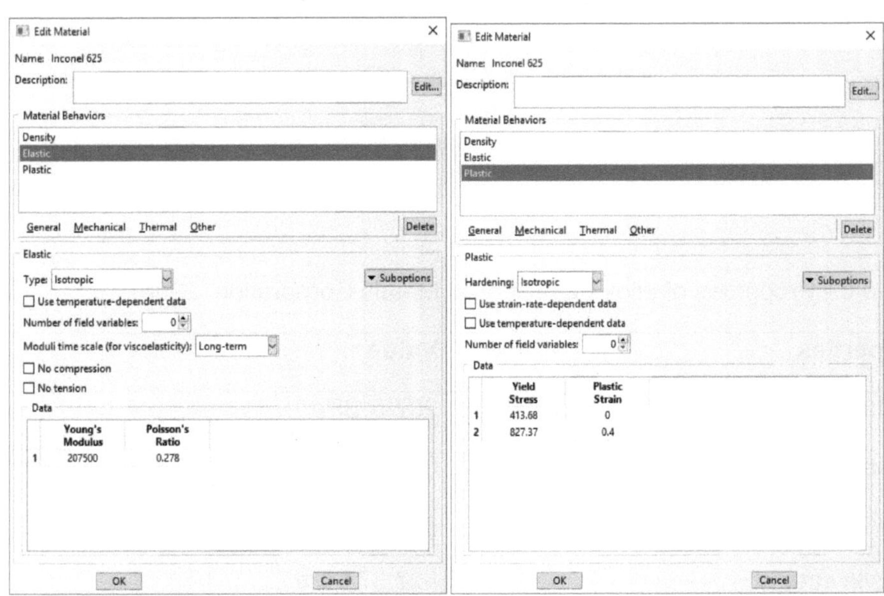

Figure 10- Entering the properties of alloy 625 into the *software.*

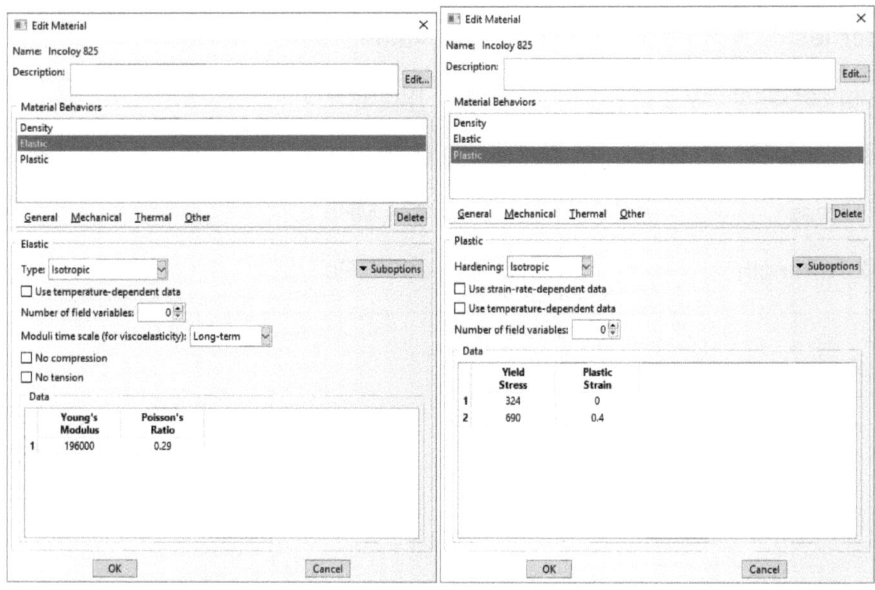

Figure 11- Entering the properties of alloy 825 into the *software.*

2.5. BOUNDARY CONDITIONS

The analysis was carried out in two stages. In the first stage, the displacement between the seats was applied, this being the energization stage.

In the second stage, internal pressure was applied.

2.5.1 Contact

In conjunction with the first stage, the contacts between the seal and the seats are initially defined, and this characteristic is later extended to the second stage as well. As tangential contact characteristics, the *Penalty-type* friction formulation was used, with a friction coefficient of 0.15. This value was found in the material prepared by (Zhao, 2011, p. 56) for the condition of static steel-steel friction coefficient without lubrication. *Hard Contact* and *Penalty* were used as normal contact characteristics. Figure 12 shows the configuration of the contact properties in the *software*.

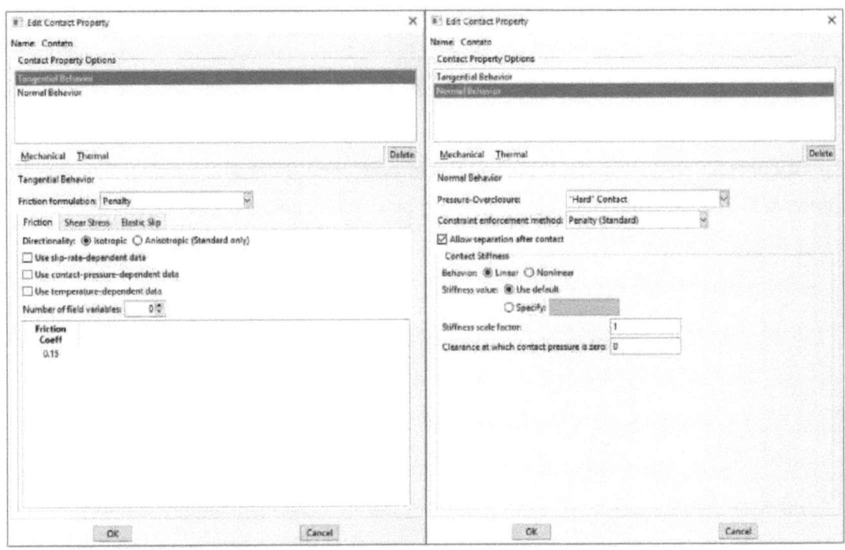

Figure 12 - Configuring contact properties in the *software*.

Figure 13 shows where the contact region for the seal and upper seat was defined. The smallest possible region was considered and defined in order to aid in a faster resolution process. Similar to Figure 13, Figure 14 shows where the contact region for the seal and lower seat was defined. The *Master* surface has been considered on the seat, while the *Slave* surface has been considered on the seal. For contacts between surfaces, it is recommended to define the *Slave surface for the* one that tends to respond to the system with greater deformations, which is why the seal surface was chosen, as the seal tends to have greater deformations in this system.

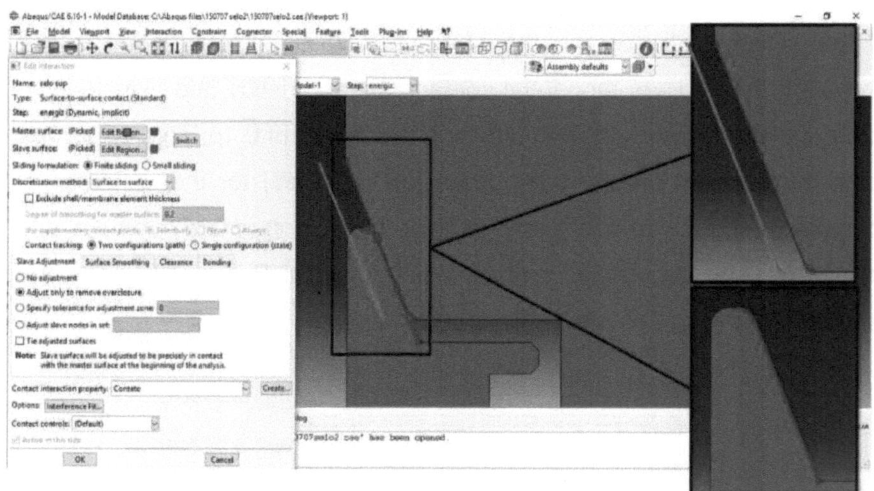

Figure 13 - Mechanical contact surface between seal and upper seat (1st model). The region shown in green is the contact region for the seat and in pink is the contact region for the seal.

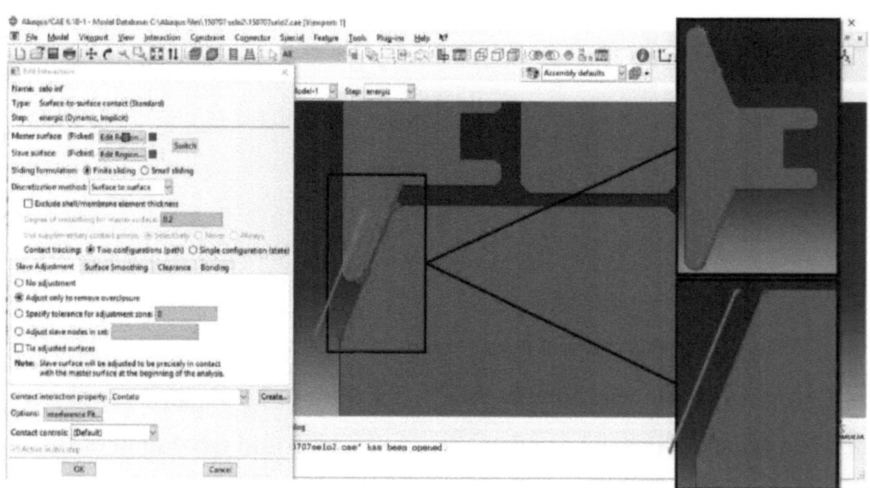

Figure 14 - Mechanical contact surface between seal and lower seat (1st model). The region shown in green is the contact region for the seat and in pink is the contact region for the seal.

2.5.2 Energization

In this first stage, the displacement between the seats was applied, with the displacement defined in Table 4. At the end of this stage, the seal is expected to be deformed by the seats in order to create a seal between the surfaces.

2.4.1. Internal pressure

In the second stage, the internal pressure of 69 MPa listed in Table 3 is applied in order to simulate the behavior under the internal working pressure. Figure 15 shows a schematic drawing of the region where the pressure was applied. Figure 16 shows the region where the pressure was applied in the ABAQUS® *software.*

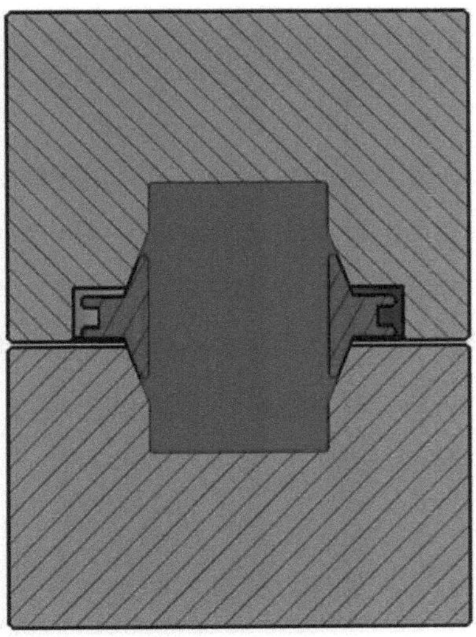

Figure 15 - Schematic drawing showing the pressure application region. The region shown in red is the internal area of the model where the pressure is applied.

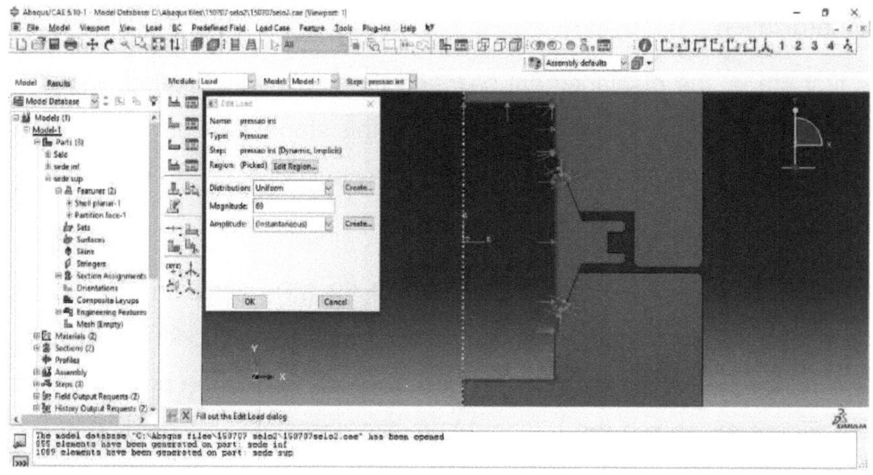

Figure 16 - Internal pressure applied to the model (1st model). The regions shown in red indicate the regions where the internal pressure was applied. The arrows in orange show the directions in which the pressure was applied.

2.6. FINITE ELEMENT MESH

With the axisymmetric model, it was decided to use fixed quadrilateral meshes. The quadrilateral mesh was configured in the *software*'s *Mesh - Controls* module (Figure 17). According to Vartziotis et al. (2013), the use of regular elements improves the quality of the mesh and, consequently, the accuracy of the solution. This is why we chose to use quadrilateral elements, as they tend to form elements with better regularization.

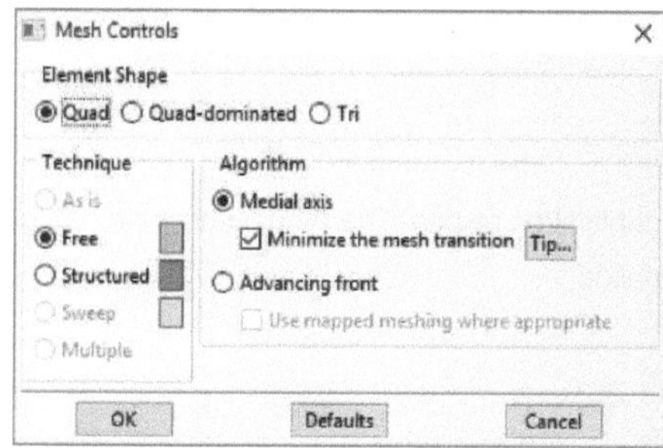

Figure 17 - Configuration of the quadrilateral mesh.

The configuration and size of the finite element mesh used is discussed in the next chapters of this paper.

2.7. SOLUTION

Once the model had been prepared, the solution method was configured in ABAQUS®. In the *Job* module, the *Job Manager* interface was used (Figure 18). When using the *Create* function, the *Edit Job* interface window appears (Figure 19). In this interface, in the *Memory* tab, the percentage of physical memory of 90% was configured, and in the *Parallelization* tab, the use of multiple processors, in this case four multiprocessors, was configured. Once the solution method has been configured, the analysis is started using the *Submit* function of the *Job Manager* interface (Figure 18). Once the analysis has been completed by the computer, the results are obtained using the *Results* function of the same interface.

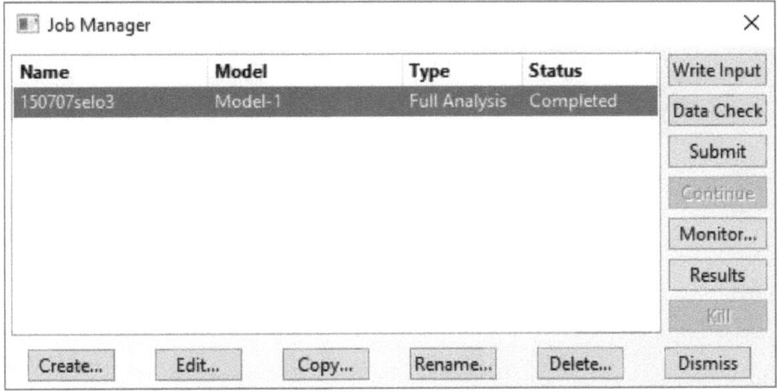

Figure 18 - *Job Manager* interface, where the jobs to be carried out are created, the analyses are submitted and the results are displayed after the analyses.

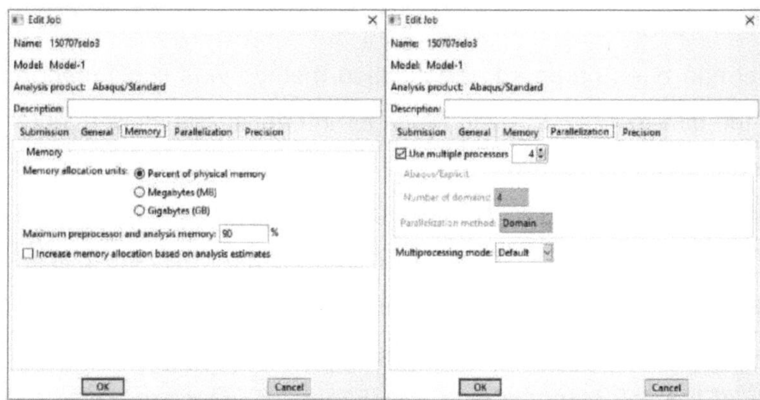

Figure 19 - *Edit Job* interface, where the analysis to be carried out is configured.

CHAPTER 3 - SIZE STUDY AND FINITE ELEMENT MESH REFINING

3.1. DEFINITIONS AND TESTS OF DIFFERENT MESH SIZES

For the first model, four different finite element mesh sizes were tested with their respective refinements in the sealing areas, in order to define which type of mesh would be used in the continuation of the other models. Table 8 shows the different mesh sizes used in the test configurations of the first model. The different mesh sizes and refinements were chosen empirically and configured in the model. The number of nodes and the number of elements in the model are the result of the mesh sizes defined.

Figures 20 and 21 show the mesh configuration for the lower seat and the mesh refinement configuration in the sealing region, for the fourth mesh configuration of the first model. The images show the area where the mesh was refined, with a greater number of elements per unit area in the sealing region of the seat, in order to find more accurate results, since this region is where the sealing strip and the contact between the seal and the seat is located. The same configuration was used for the upper seat as for the lower seat.

Table 8 - Mesh test configurations used in the first model.

Configuration	Mesh size on seal	Size of mesh refinement on the seal	Mesh size on the seat	Mesh refinement size at headquarters	Number of nodes in the model	Number of model elements
First Configuration	12	2,5	30	5	544	450
Second configuration	10	1,0	20	2	860	736
Third	7	0,7	12	1,5	1.161	1.016

Configuration						
Wednesday Configuration	5	0,5	8	1	3.041	2.804

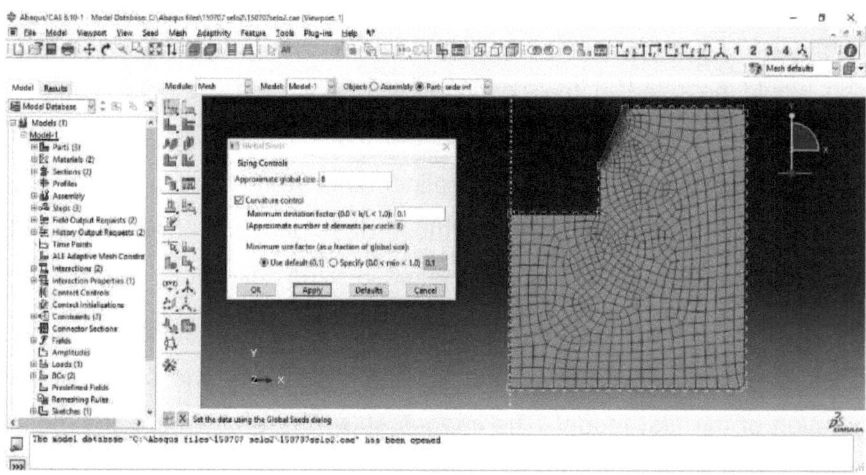

Figure 20 - General finite element mesh configuration for the lower seat, for the fourth configuration of the first model.

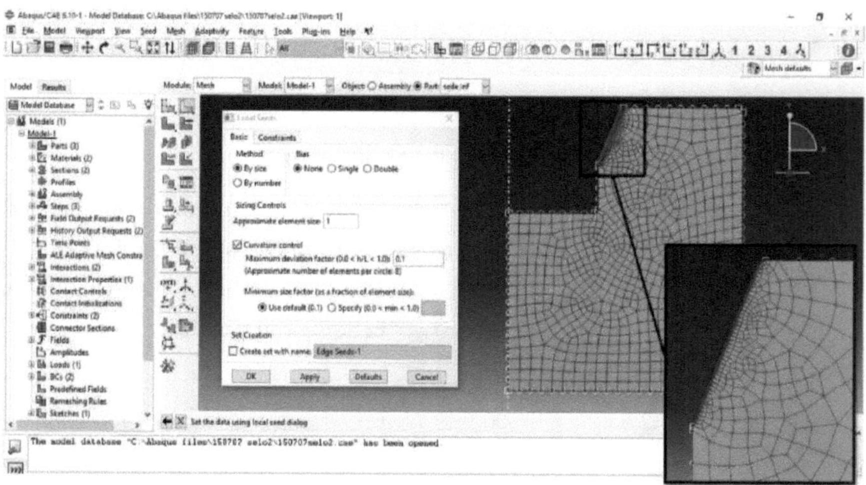

Figure 21 - Finite element mesh refinement configuration in the lower seat sealing region, for the fourth configuration of the first model. The region to which the refinement was applied is highlighted in red. It can be seen that after generating the mesh, it is smaller than the rest of the model.

Figures 22 and 23 show the mesh configuration for the seal and the mesh refinement configuration in the sealing region, for the fourth mesh configuration of the first model. In the images, it is possible to see in the area where the mesh was refined, a greater number of elements per unit area in the sealing region of the seal, in order to find more accurate results, since this region is where the sealing strip and the contact between the seal and the seat is located.

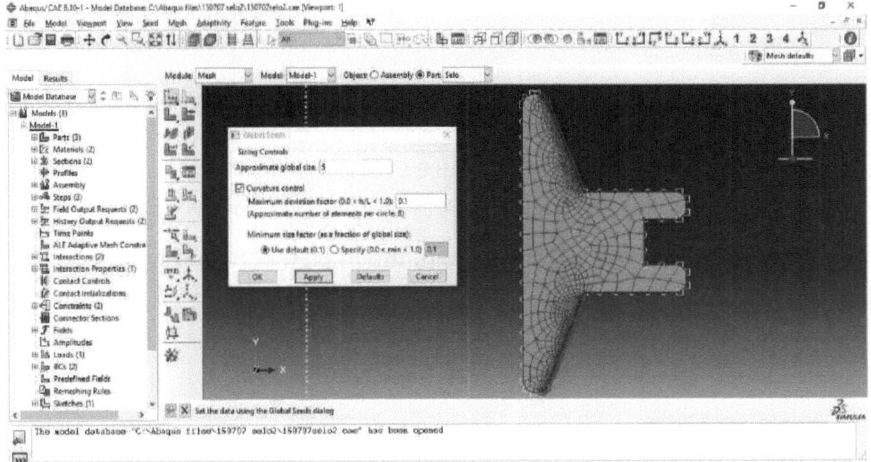

Figure 22 - General finite element mesh configuration for the seal, for the fourth configuration of the first model.

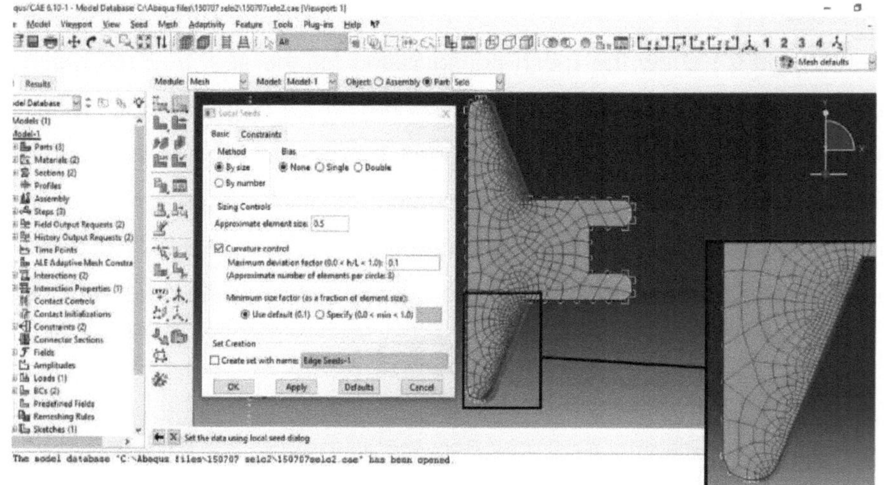

Figure 23 - Finite element mesh refinement configuration in the seal sealing region, for the fourth configuration of the first model. The region to which the refinement was applied is highlighted in red. It can be seen that after the mesh has been generated, it is smaller than the rest of the model.

Figure 24 shows the different finite element mesh size configurations used in the first model.

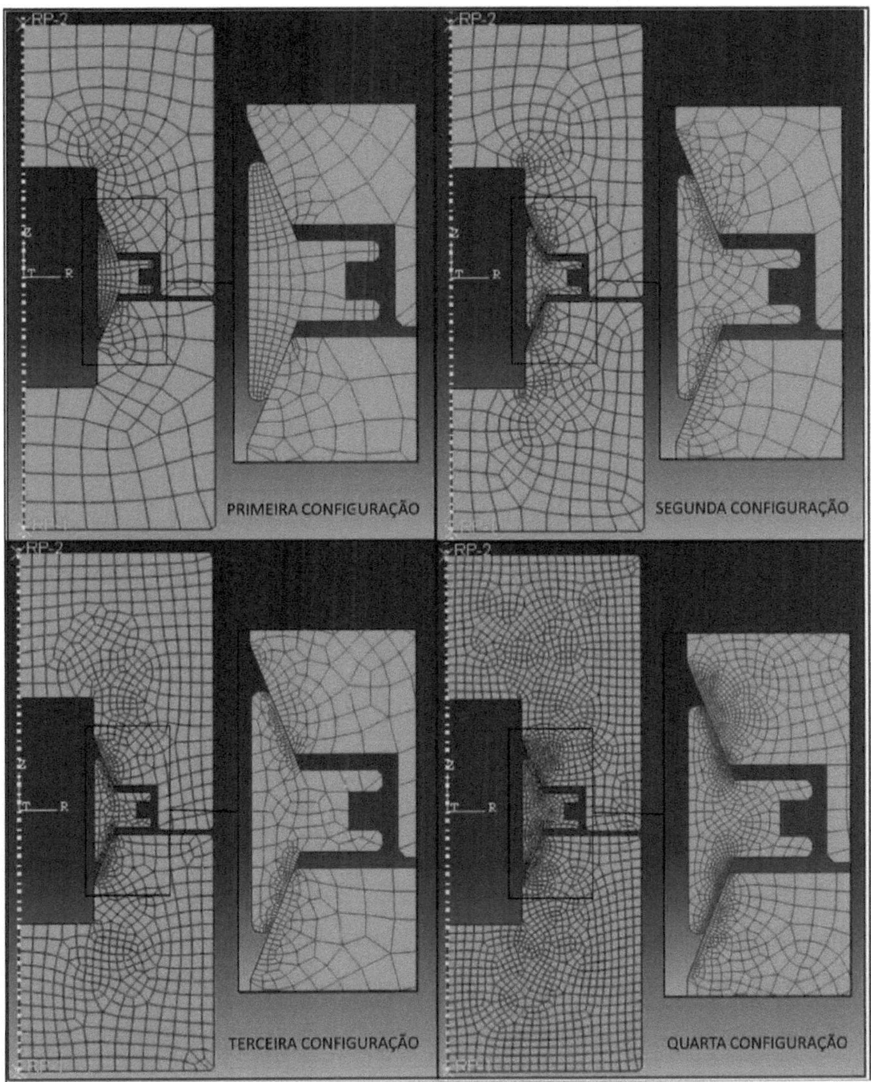

Figure 24 - Finite element mesh configurations used for the first model. The first configuration shows the mesh used with the least refinement. The fourth configuration shows the mesh used with greater refinement.

3.2. RESULTS OBTAINED IN THE STUDY OF FINITE ELEMENT MESH SIZE AND REFINEMENT

In the study of finite element mesh size and refinement, the results after the final stage were checked for:

- von Mises tension;

- Plastic deformation;

- Contact pressure.

Table 9 shows a summary of the maximum values found for the von Mises stress, plastic deformation and contact pressure, as well as the calculation execution time and the number of iterations for the four finite element mesh configurations used as a test in the first model after the final stage, used to define which configuration would be used to continue the study for the other models developed in the seal optimization study. Figures 25 to 28 show the maximum values achieved in graph form.

Figure 25 - Graph of the von Mises stress after the final stage, for the four mesh size configurations - Maximum values.

Figure 25 shows the maximum values found for the von Mises stress for the four finite element mesh configurations used as a test in the first model after the final stage. It can be seen that the maximum value of the von Mises stress in the seal tends to increase as the mesh refinement increases, while the values in the seats show little variation.

Table 9 - Results after the final step (maximum values), calculation execution time and number of iterations for the finite element mesh size test.

Model	von Mises stress [MPa]		Plastic deformation [%]		Contact pressure [MPa]	Calculation execution time	Number of iterations
	Seal	Headquarters	Seal	Headquarters			

						(seconds)	
First configuration	338,0	408,5	3,187	0,145	588,4	24	172
Second configuration	360,7	413,8	3,864	0,322	863,3	30	212
Third configuration	366,5	413,9	4,665	0,093	885,3	35	250
Fourth Configuration	409,1	413,5	9,494	0,196	991,0	63	290

Figure 26 - Graph of Plastic Deformation (%) after the final stage, for the four mesh size configurations - Maximum values.

Figure 26 shows the maximum values found for plastic deformation for the four finite element mesh configurations used as a test in the first model after the final stage. It can be seen that the maximum plastic deformation value in the seal tends to increase as the mesh refinement increases, while the values in the seats vary unevenly.

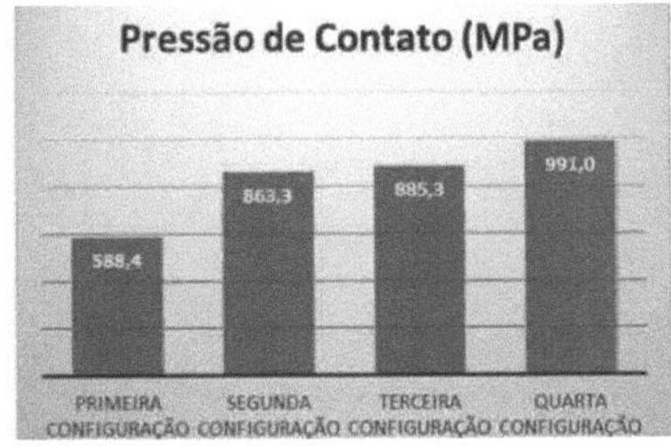

Figura 27 - Graph of the Contact Pressure (MPa) after the final stage, for the four configurations

41

mesh size configurations - Maximum values.

Figure 27 shows the maximum values found for the contact pressure for the four finite element mesh configurations used as a test in the first model after the final stage. It can be seen that the maximum contact pressure value tends to increase as the mesh refinement increases.

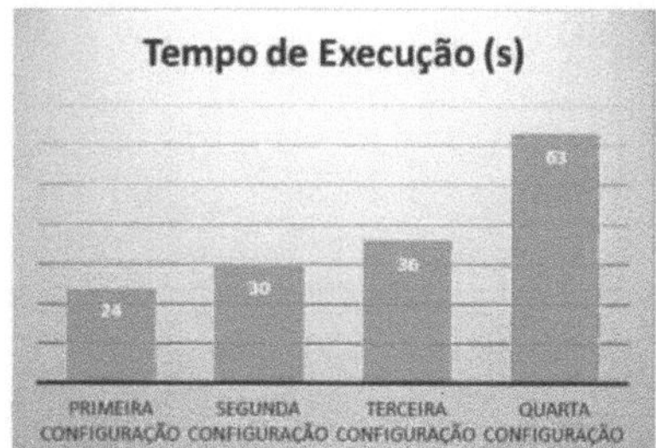

Figura 28 - Graph of execution time (s) for the four mesh size configurations.

Figure 28 shows the calculation execution time for the four finite element mesh configurations used as a test in the first model. The increase in execution time due to the increase in refinement is justified because the greater the refinement, the greater the number of nodes and elements in the model that must be calculated by the *software*.

Figures 29 to 34 show the results for von Mises stress, plastic deformation and contact pressure for the four finite element mesh configurations used as a test in the first model after the final stage.

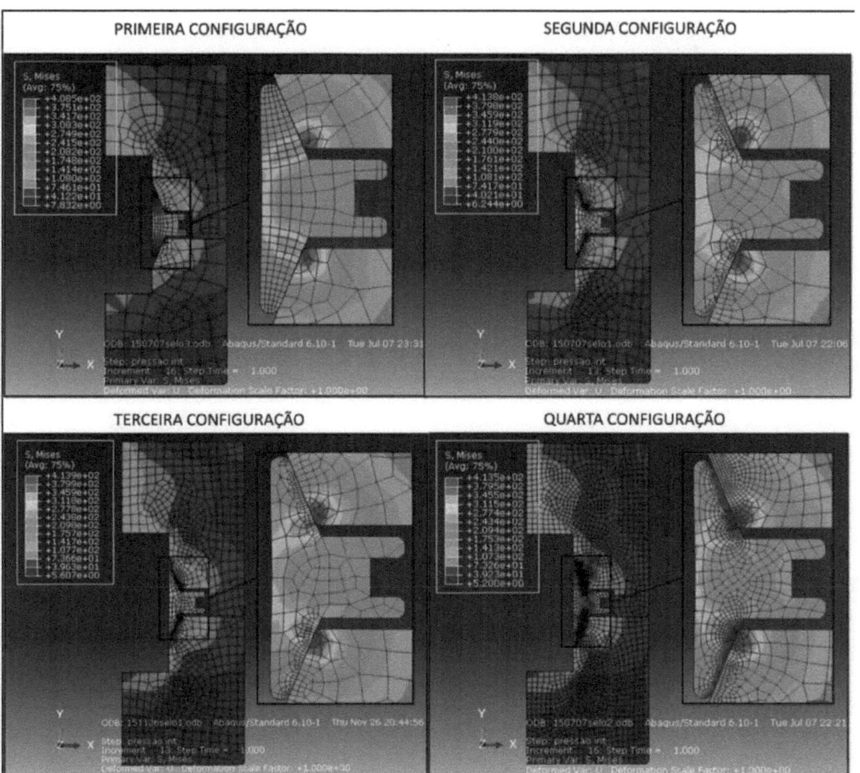

Figure 29 - Von Mises stress (MPa) for the four mesh size configurations applied to the first model.

Figure 29 shows the results for the von Mises stress for the four mesh size configurations applied to the first model. It can be seen that the general behavior of the system remains the same regardless of the mesh size, as do the maximum stress values.

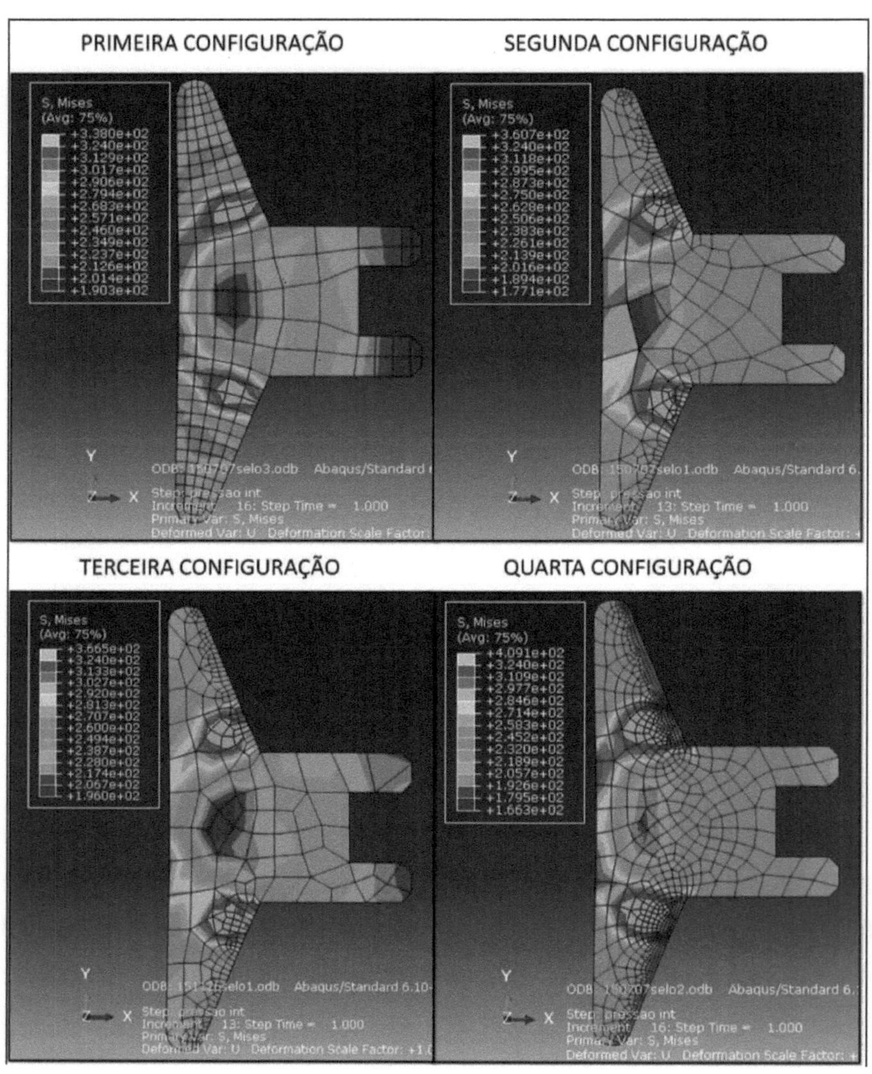

Figure 30 - Von Mises stress (MPa), for the four mesh size configurations for the seal. The stress value filter was applied to the yield strength of the material. The regions shown in gray exceed the yield strength.

Figure 30 shows the results for the von Mises stress for the four mesh size configurations applied to the first seal model. It can be seen that the general behavior of the seal remains the same regardless of the mesh size, but the maximum stress value tends to increase due to the greater refinement of the mesh.

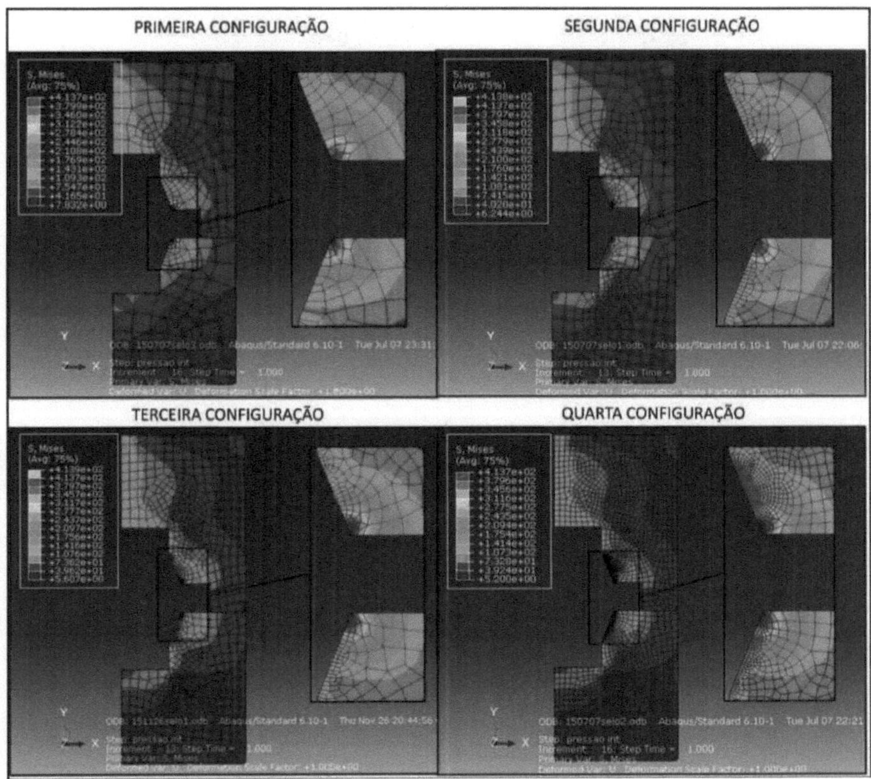

Figure 31 - von Mises stress (MPa), for the four mesh size configurations for the seats. The stress value filter was applied to the yield stress of the material.

Figure 31 shows the results for the von Mises stress for the four mesh size configurations applied to the first seat model. It can be seen that the general behavior of the seats remains the same regardless of the mesh size, as do the maximum stress values.

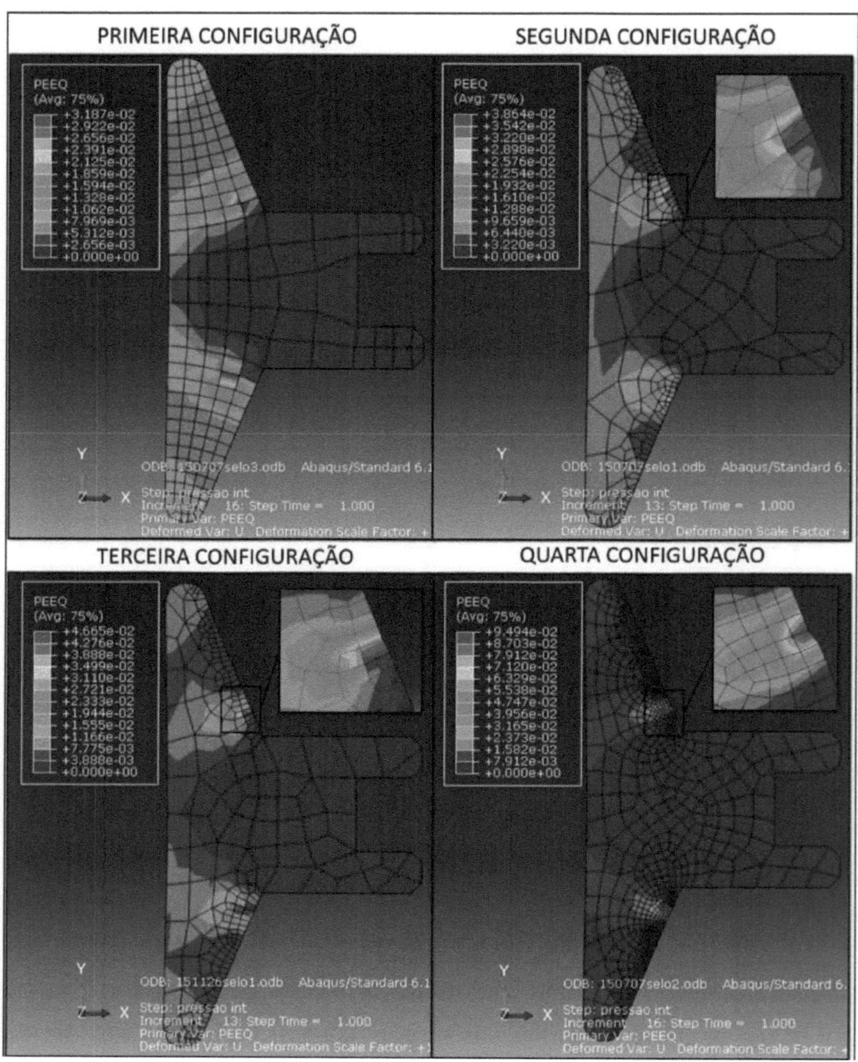

Figura 32 - Plastic deformation (%), for the four mesh size configurations for the
seals.

Figure 32 shows the plastic deformation results for the four mesh size configurations applied to the first seal model. It can be seen that the general behavior of the seal remains the same regardless of the mesh size, but the maximum value of plastic deformation tends to increase due to the greater refinement of the mesh.

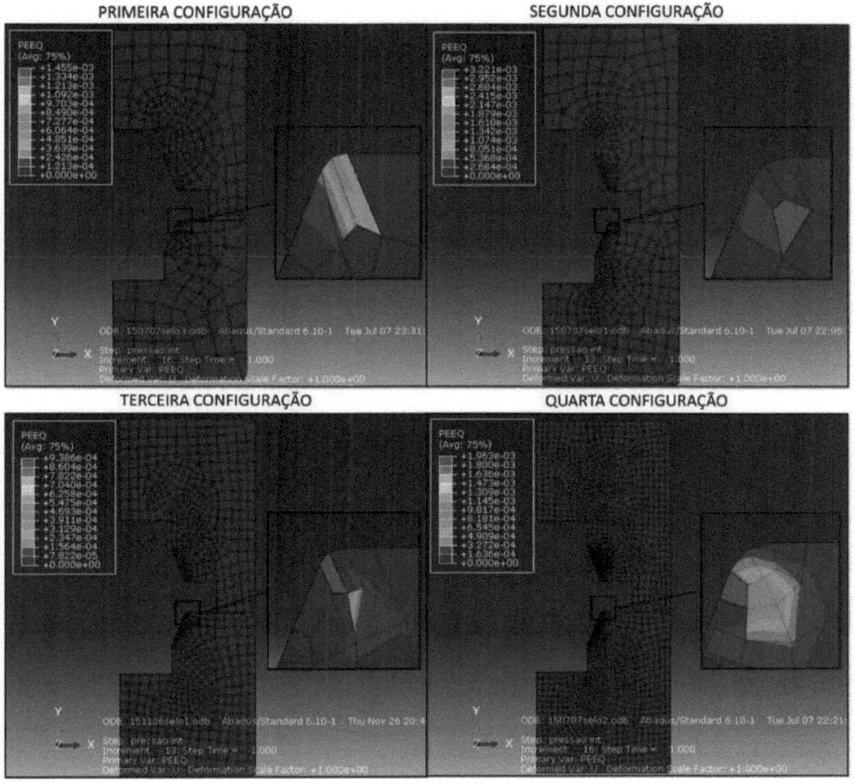

Figura 33   - Plastic deformation (%), for the four mesh size configurations for the seats.

Figure 33 shows the plastic deformation results for the four mesh size configurations applied to the first seat model. It can be seen that the general behavior of the seats remains the same regardless of the mesh size. For all four configurations, only a small region showed plastic deformation in the seats.

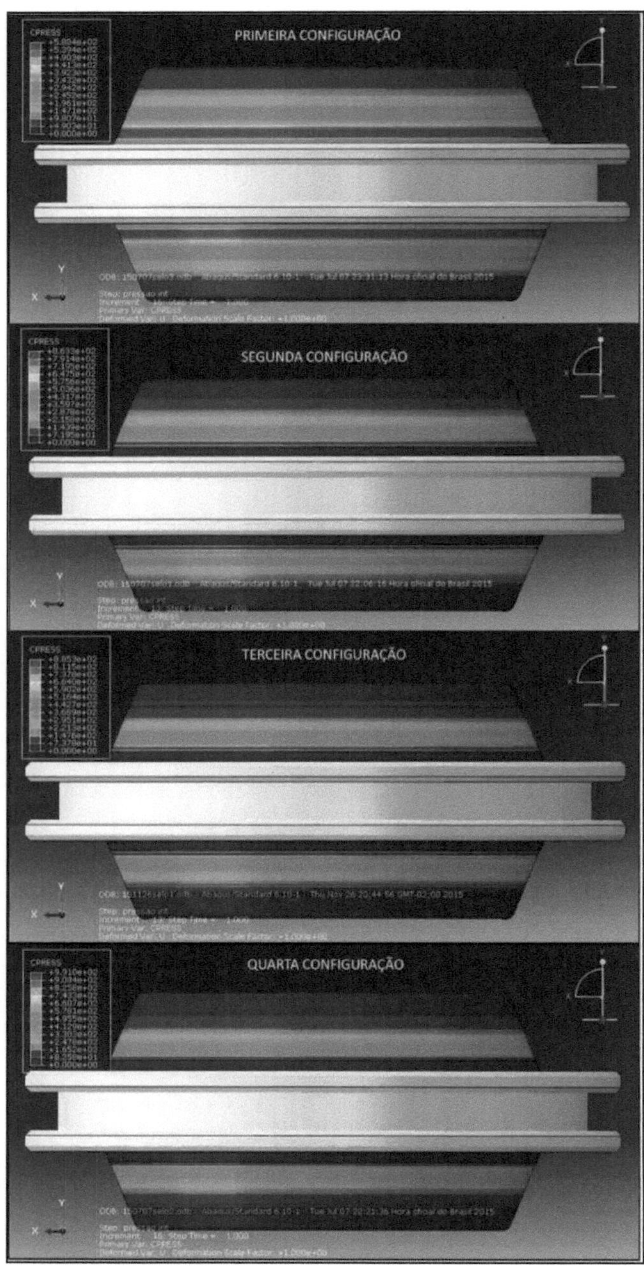

Figure 34 - Contact pressure (MPa) for the four mesh size configurations.

Figure 34 shows the contact stress results for the four mesh size configurations applied to the first model for the seal. For better visualization, the axisymmetric profile has been revolutionized by 360°. It can be seen that the general behavior of the seal remains the same regardless of the mesh size, but the maximum contact stress tends to increase due to the greater refinement of the mesh. It is also possible to verify the location of the effective sealing range with greater precision due to the greater refinement.

3.3. ANALYSIS OF THE RESULTS OF THE FINITE ELEMENT MESH STUDY

In the finite element mesh study, it can be seen in Figures 29 to 34 that the general behavior of the system remains the same, regardless of the mesh refinement used. However, the results for the maximum values of von Mises stress, plastic deformation and contact pressure tend to increase. This confirms what was described by Manzoli et al. (2014), which states that the more refined a finite element mesh is, the more it tends to show more accurate results.

After analyzing the results, due to the relatively short time it took to carry out the analysis, regardless of the types of mesh presented, we decided to use the fourth option of finite element mesh for the other models developed in the continuation of the study, since this is the most refined mesh among the four developed, tending to present more accurate results.

3.4. FINITE ELEMENT MESH SIZING AND REFINING USED TO DEVELOP AND OPTIMIZE MODELS

After defining the mesh size and refinement to be used, similar meshes were created for each model. Figures 20 to 23 above show the meshes used for the entire sealing system in the fourth configuration of the first model, which was considered the most appropriate configuration for this study. For the other models, similar meshes using the same configuration were used. Table 10 shows the number of nodes and elements used for each model.

Table 10 - Number of nodes and elements for each model.

Model	Number of nodes	Number of elements
First model	3.041	2.804
Second model	3.035	2.798
Third model	2.986	2.749
Model Room	3.077	2.850
Fifth Model	2.935	2.713
Sixth Model	2.933	2.710

Table 10 shows the number of nodes and elements for each model. In this table, you can see how refined the mesh is for the whole set.

CHAPTER 4 - RESULTS AND DISCUSSION

4.1. RESULTS OBTAINED

For the main study in this paper, which refers to the model optimization process, the results of the system were checked after the two stages:

- Energization;

- Application of internal pressure (final stage).

The verification after energization was checked, as the von Mises stresses generally reach higher values than in the final stage. In addition, it is possible to check that after energization the system will present acceptable values, which tend to maintain this condition after the internal pressure application stage. The verification after applying internal pressure is the most important phase. With these results, it is possible to check whether the design tends to respond with acceptable values when operated under the expected working conditions.

The following results were checked for each stage:

- von Mises tension;

- Plastic deformation;

- Contact pressure.

The von Mises stress is evaluated in order to check how, where and if the seal has exceeded the yield stress of the material and how close this stress has come to the material's tensile strength. Plastic deformation is evaluated in order to check if and where permanent deformation has occurred in the system after it has been energized and internal pressure has been applied. The most important result is the contact pressure. This is assessed in order to check whether the system has contact pressure between the seal and the seats after the final stage.

4.1.1. Results After Energization

Table 11 shows a summary of the maximum values found for von Mises stress, plastic deformation and contact pressure for each model at the end of the first stage.

Table 11 - Results after the energization stage (maximum values).

Model	Von voltage Mises [MPa]		Plastic deformation [%]		Contact pressure [MPa]
	Seal	Headquarters	Seal	Headquarters	
First Model	406,1	413,8	8,978	0,196	1112,6
Second Model	347,5	201,7	2,569	0	269,0
Third Model	336,6	165,9	1,381	0	218,2
Model Room	337,9	233,9	2,078	0	333,8
Fifth Model	344,8	231,2	2,340	0	385,0
Sixth Model	335,4	294,1	1,339	0	593,7

Table 11 shows the results after the system is energized and before the internal pressure is applied. The desired result for this system is for it to have a contact pressure value greater than the internal pressure value applied to the system. All the models have values in excess of the internal pressure to be applied of 69 MPa. The first model has a contact pressure 16.1 times greater than the expected internal pressure, the second model 3.9 times, the third model 3.2 times, the fourth model 4.8 times, the fifth model 5.6 times and the sixth model 8.6 times.

Figures 35 to 38 show the maximum values achieved in graph form, after the energization stage and before the internal pressurization stage.

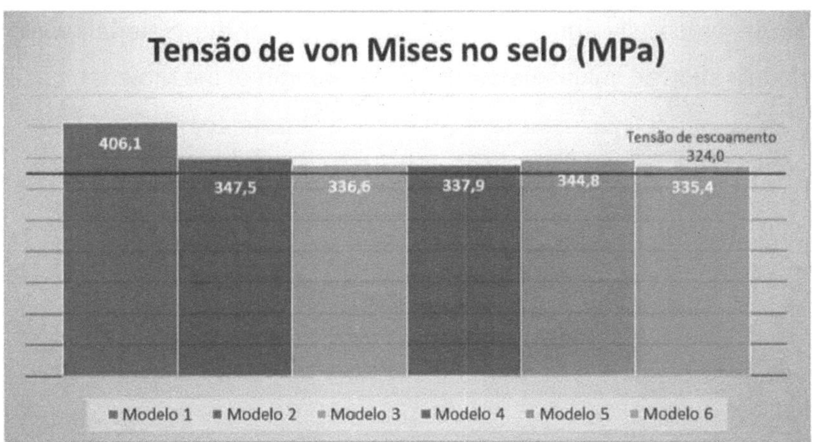

Figure 35 - Graph of the von Mises stress (MPa) in the seal after the energization stage, for the six models developed - Maximum values.

Figure 35 shows the maximum values found for the von Mises stress in the seal for each model at the end of the first stage. All the models showed the maximum von Mises stress value exceeding the seal's yield stress value. While models 2 to 6 show little variation and very close values, model 1 has a significantly higher maximum value than the others.

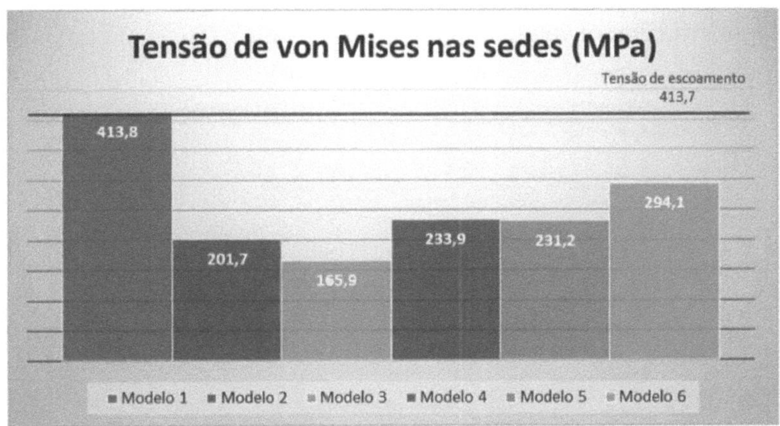

Figure 36 - Graph of the von Mises stress (MPa) in the seats after the energization stage, for the six models developed - Maximum values.

Figure 36 shows the maximum values found for the von Mises stress in the seats for each model at the end of the first stage. The first model showed a maximum value practically equal to the yield strength of the material, while the other models showed values below the yield strength of the material.

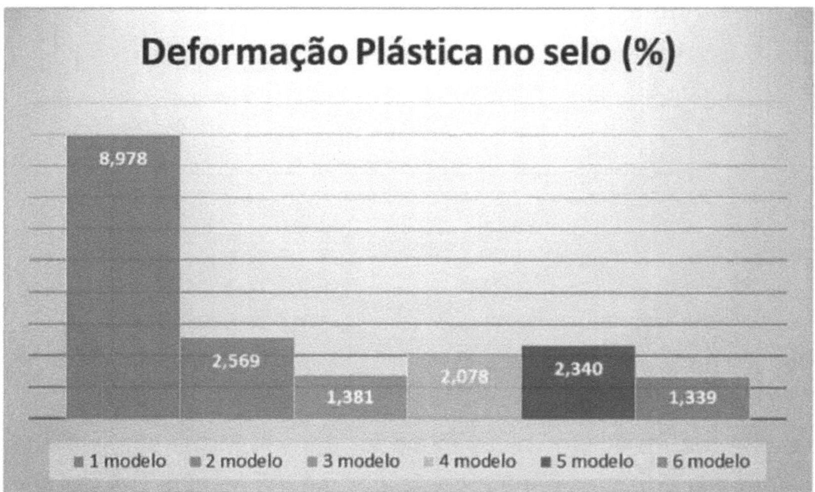

Figure 37 - Graph of plastic deformation (%) in the seal after the energizing stage, for the six models developed - Maximum values.

Figure 37 shows the maximum values found for plastic deformation in the seal for each model at the end of the first stage. The first model had a significantly higher maximum value than the other models.

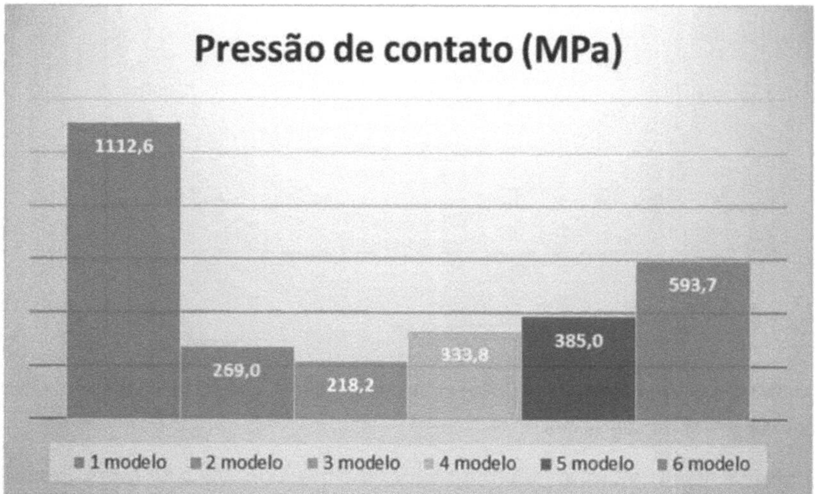

Figure 38 - Graph of the contact pressure (MPa) after the energization stage, for the six models

54

developed - Maximum values.

Figure 38 shows the maximum values found for the contact stress for each model at the end of the first stage. The first model had a significantly higher maximum value than the other models.

Figures 39 to 43 show the results for von Mises stress, plastic deformation and contact pressure for the six finite element models developed in the optimization phase after the energization stage and before the internal pressurization stage.

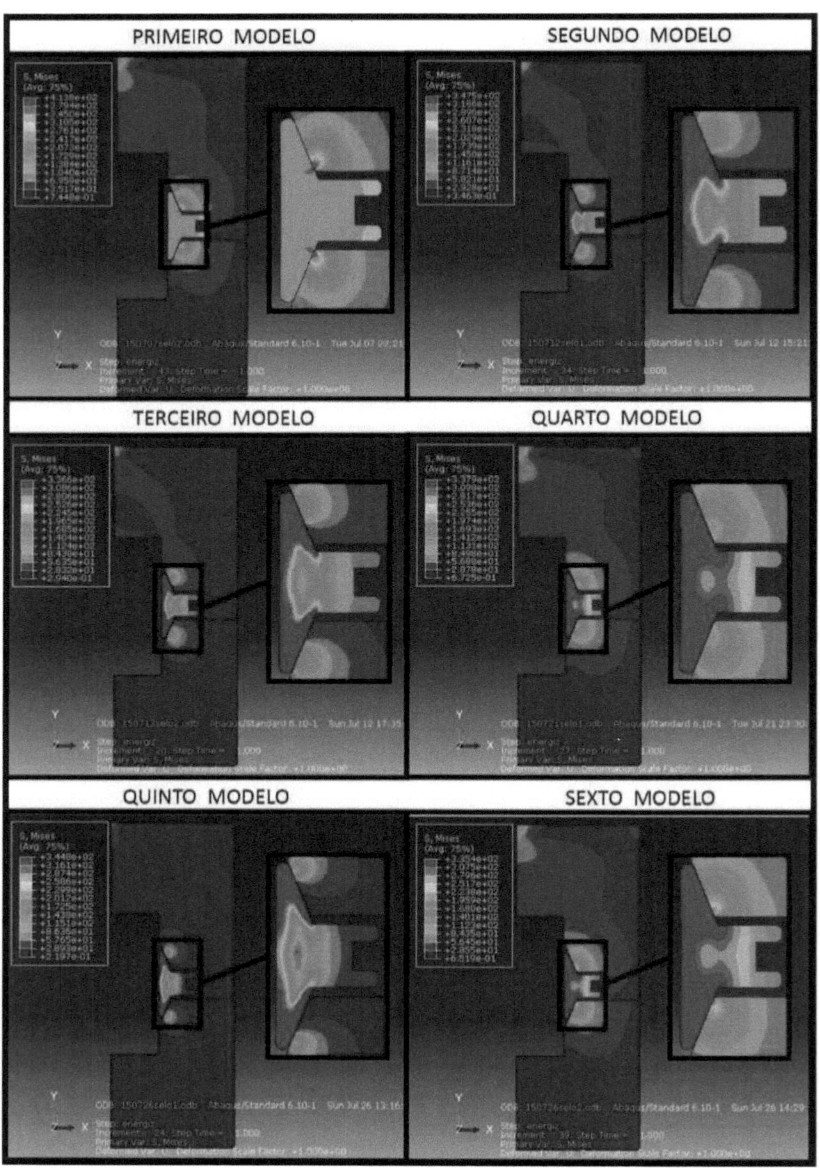

Figura 39 - von Mises stress for the six models developed after energization [MPa].
It is possible to check the stresses throughout the system. The closer to the sealing regions, the higher the stresses.

Figure 39 shows the results for the von Mises stress for each model at the end of the first stage. In all the models, it is possible to see the increase in stress the closer you get to the sealing regions.

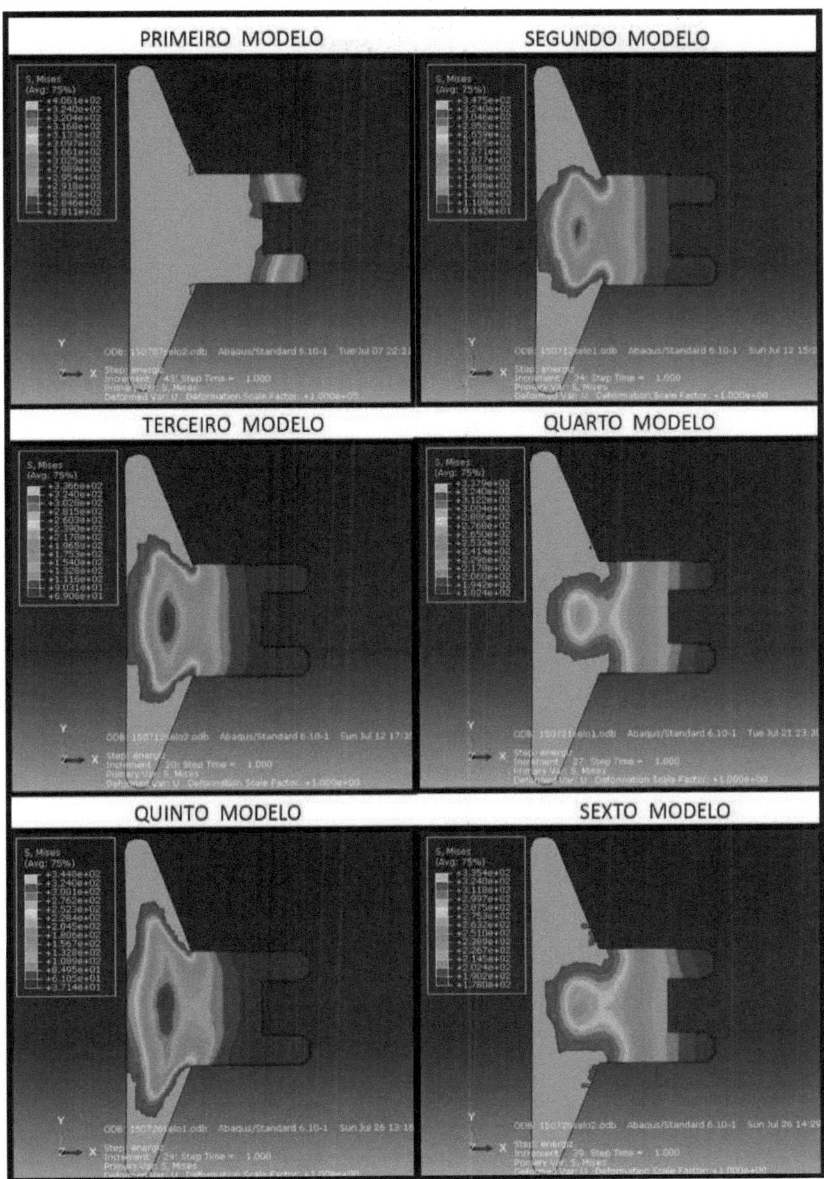

Figura 40 - von Mises stress in the seal for the six models developed after energization [MPa]. The filter was applied to the seal's yield stress value. The regions shown in gray exceed the material's yield stress values.

Figure 40 shows the results for the von Mises stress in the seal for each model at the end of the first stage. The regions shown in gray exceed the yield strength of the material. The first model exceeded the yield stress almost entirely, while the others showed regions with stress above the yield limit at the ends of the seal, close to the effective sealing area.

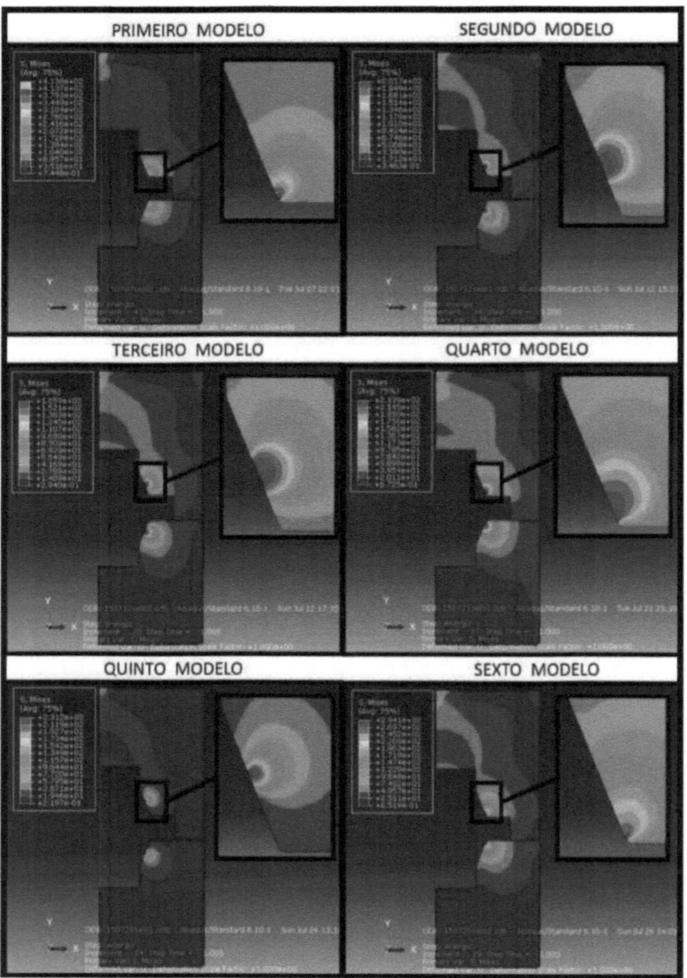

Figure 41 - Von Mises stress in the seats for the six models developed after *energization [MPa]. The filter was only applied to the first model, which showed values that exceeded the yield strength of the material. The regions shown in gray exceed the material's yield strength values.*

Figure 41 shows the results for the von Mises stress in the seats for each model at the end of the first stage. All the models showed an increase in stress in the areas around and in the sealing strip. The first model showed a maximum stress value in the sealing region practically equal to the yield stress value of the material, represented in gray in the figure.

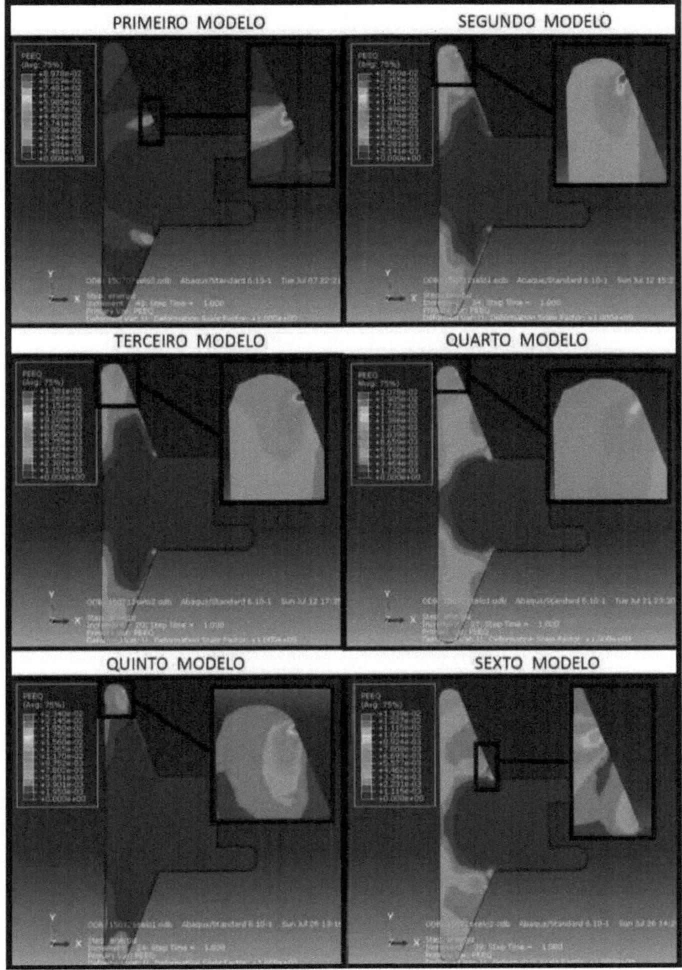

Figure 42 - Plastic deformation in the seal for the six models developed after energization [%]. The blue regions showed no plastic deformation. Approaching the effective sealing range of each model, the permanent deformation values tend to show maximum values.

Figure 42 shows the results for plastic deformation in the seal for each model at the end of the first stage. All the models showed an increase in plastic deformation in the areas around and in the sealing strip. The first model had a significantly higher maximum value than the other models.

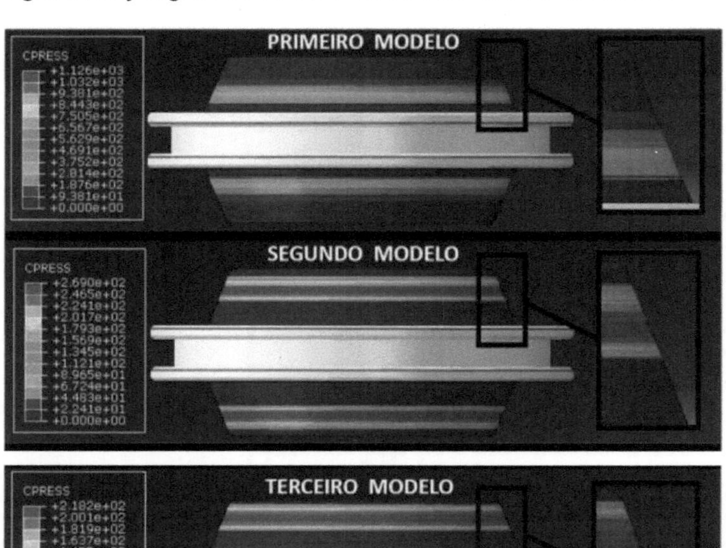

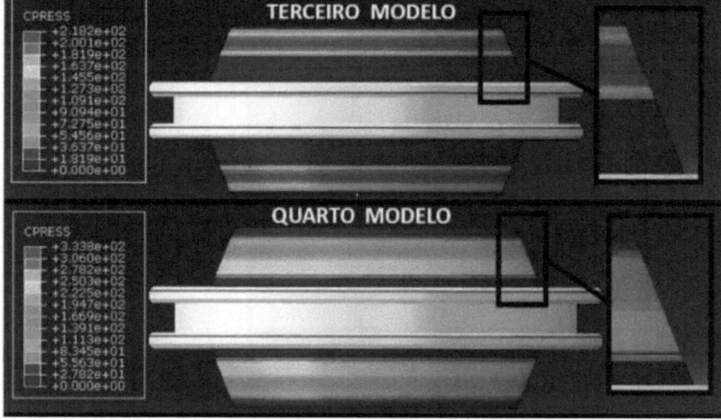

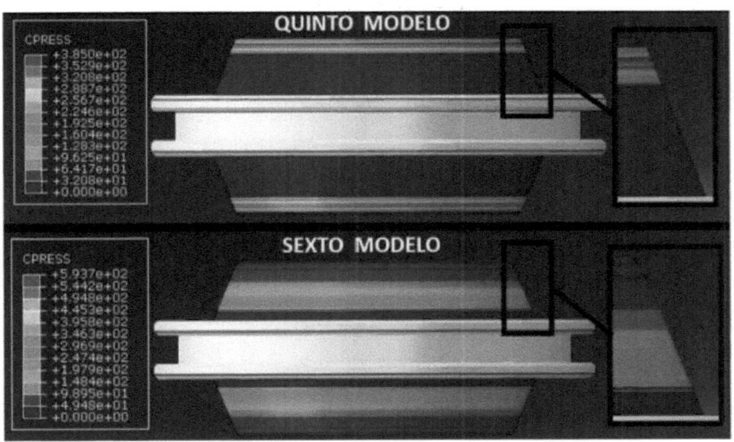

Figure 43 - Contact pressure for the six models developed, after energization [MPa]. The regions shown in red are the regions where the maximum contact pressure values of the seal are shown. These are the regions considered to be the effective sealing range.

Figure 43 shows the results for the contact stress for each model at the end of the first stage. The first model had a significantly higher maximum value than the other models. The region shown in red in each model is the area with the highest contact tension, which is the effective sealing area in the seal.

4.1.2. Results obtained after the final stage

Table 12 shows a summary of the maximum values found for von Mises stress, plastic deformation, contact pressure and calculation execution time for each model after the final stage.

Table 12 - Results after the final stage (maximum values).

Model	von Mises stress [MPa]		Plastic deformation [%]		Contact pressure [MPa]	Calculation execution time [s]	Number of iterations
	Seal	Headquarters	Seal	Headquarters			
First	409,	413,5	9,49	0,196	991,0	63	290

Model	1		4				
Second Model	330,3	292,6	2,569	0	340,7	52	241
Third Model	326,6	258,8	1,381	0	293,4	69	153
Bedroom Model	329,8	329,9	2,246	0	344,5	56	247
Fifth Model	326,7	281,3	2,340	0	418,8	52	197
Sixth Model	337,9	366,8	1,686	0	534,4	75	271

The results shown in Table 12 show that all the models have contact pressure after the system has been energized and internal pressure has been applied, which is the basic premise for ensuring the system is sealed. As the internal pressure applied is 69 MPa, the models have a sealing coefficient in the order of 14.6 times for the first model, 4.9 times for the second model, 4.3 times for the third model, 5.0 times for the fourth model, 6.1 times for the fifth model and 7.7 times for the sixth model.

Figures 44 to 47 show the maximum values achieved in graph form after the final stage.

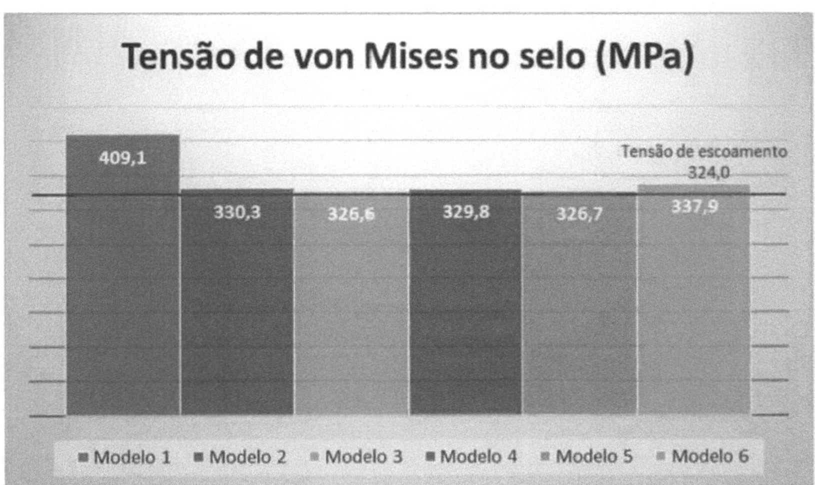

Figure 44 - Graph of the von Mises stress (MPa) in the seal after the final stage, for the six models developed - Maximum values.

Figure 44 shows the maximum values found for the von Mises stress in the seal for each model after the final stage. All the models showed the maximum von Mises stress value exceeding the seal's yield stress value. While models 2 to 6 show little variation and very close values, model 1 has a significantly higher maximum value than the others. Compared to Figure 35, which shows the maximum von Mises stress values in the seal after the energization stage, there is a drop in values in models 2 to 6. Model 1 shows an insignificant increase in its maximum value.

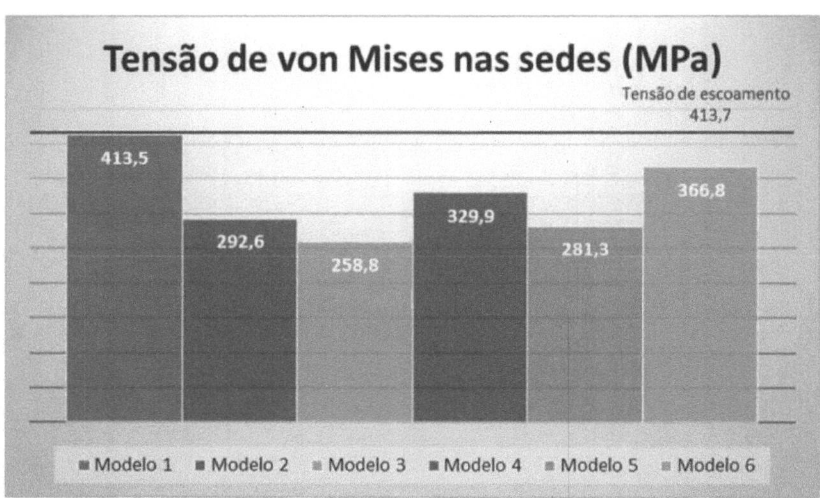

Figure 45 - Graph of the von Mises stress (MPa) in the seats after the final stage, for the six models developed - Maximum values.

Figure 45 shows the maximum values found for the von Mises stress in the seats for each model after the final stage. The first model showed a maximum value practically equal to the yield strength of the material, while the other models showed values below the yield strength of the material. Compared to Figure 36, which shows the maximum von Mises stress values in the seats after the energization stage, there is an increase in the values in models 2 to 6. Model 1 shows an insignificant reduction in its maximum value.

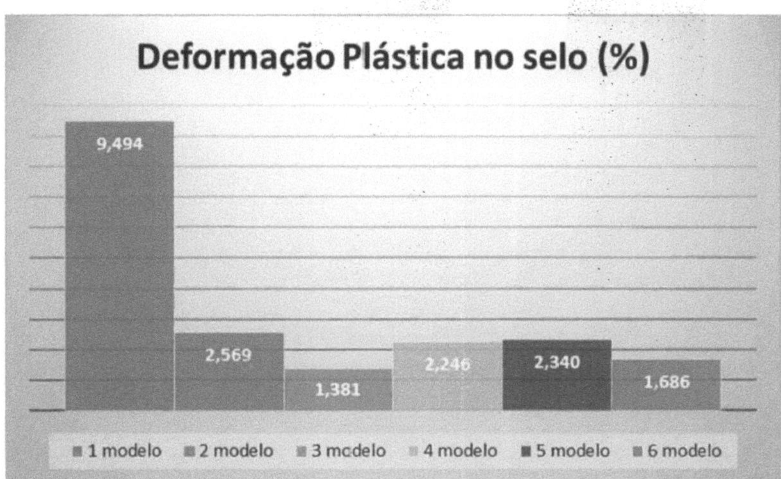

Figure 46 - Graph of plastic deformation (%) in the seal after the final stage, for the six models developed - Maximum values.

Figure 46 shows the maximum values found for plastic deformation in the seal for each model after the final stage. The first model showed a significantly higher maximum value than the other models. Compared with Figure 37, which shows the maximum values found for plastic deformation in the seal after the energization stage, it can be seen that models 1, 4 and 6 showed a slight increase in their values, while the other models remained with their maximum values unchanged.

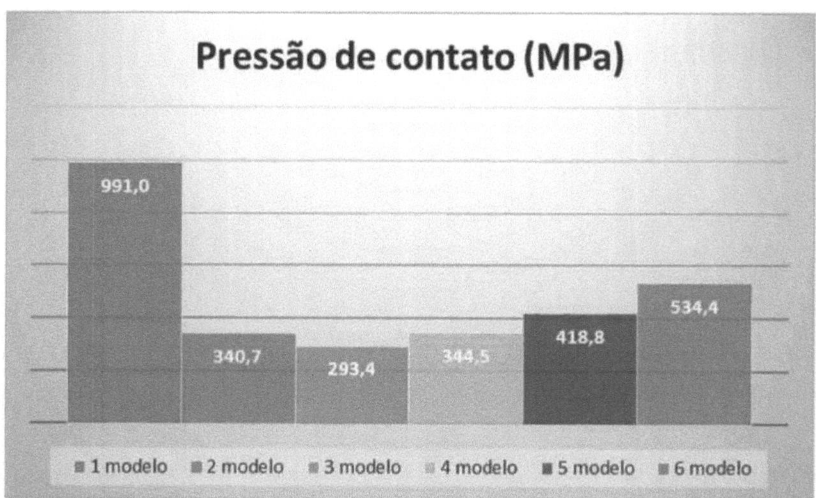

Figure 47 - Graph of the contact pressure (MPa) after the final stage, for the six models developed - Maximum values.

Figure 47 shows the maximum values found for the contact stress for each model after the final stage. The first model showed a significantly higher maximum value than the other models. Compared to Figure 38, which shows the maximum values found for the contact voltage after the energization stage, models 1 and 6 showed a reduction in their values, while the other models showed an increase.

Figures 48 to 52 show the results for von Mises stress, plastic deformation and contact pressure for the six finite element models developed in the optimization phase after the final stage.

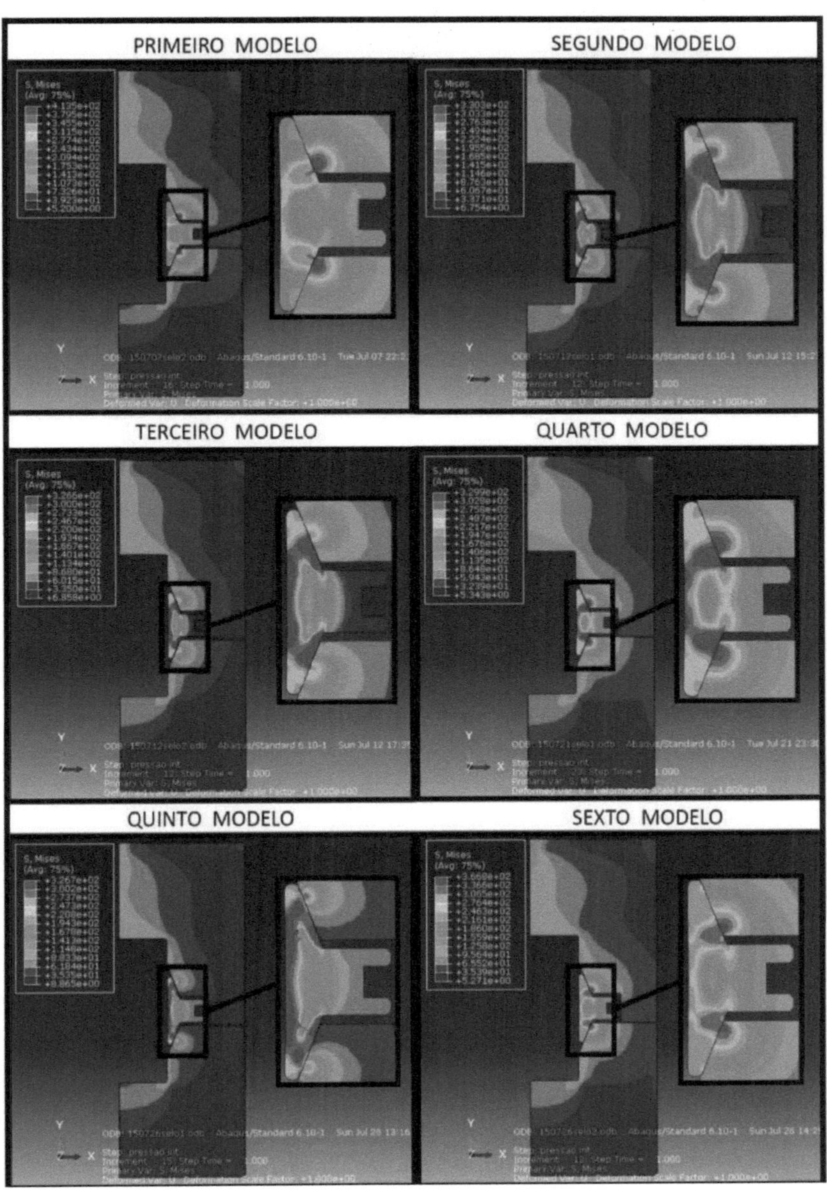

Figure 48 - von Mises stress for the six models developed after the final stage [MPa].

it is possible to check the stresses throughout the system. The closer to the sealing regions, the higher the stresses.

Figure 48 shows the results for the von Mises stress for each model after the final stage. In all the models, it is possible to see the increase in stress the closer the sealing regions are.

Compared to Figure 39, where the von Mises results are shown after the energization stage, it can be seen that after the final stage the models show an increase in stress in the areas close to the application of internal pressure, in addition to maintaining stress levels in the sealing region close to the levels shown in the previous stage. However, it can be seen that the highest tensions remain close to the sealing areas.

Also, at this stage, it is easier to see the location of the sealing strip in each seal, due to the increase in tension near the sealing areas: the first, fourth and sixth models have a sealing strip located in the center of the seal; the second and third models have a sealing strip in the middle of the seal, i.e. between the center and the end of the seal; the fifth model has a sealing strip at the end of the seal.

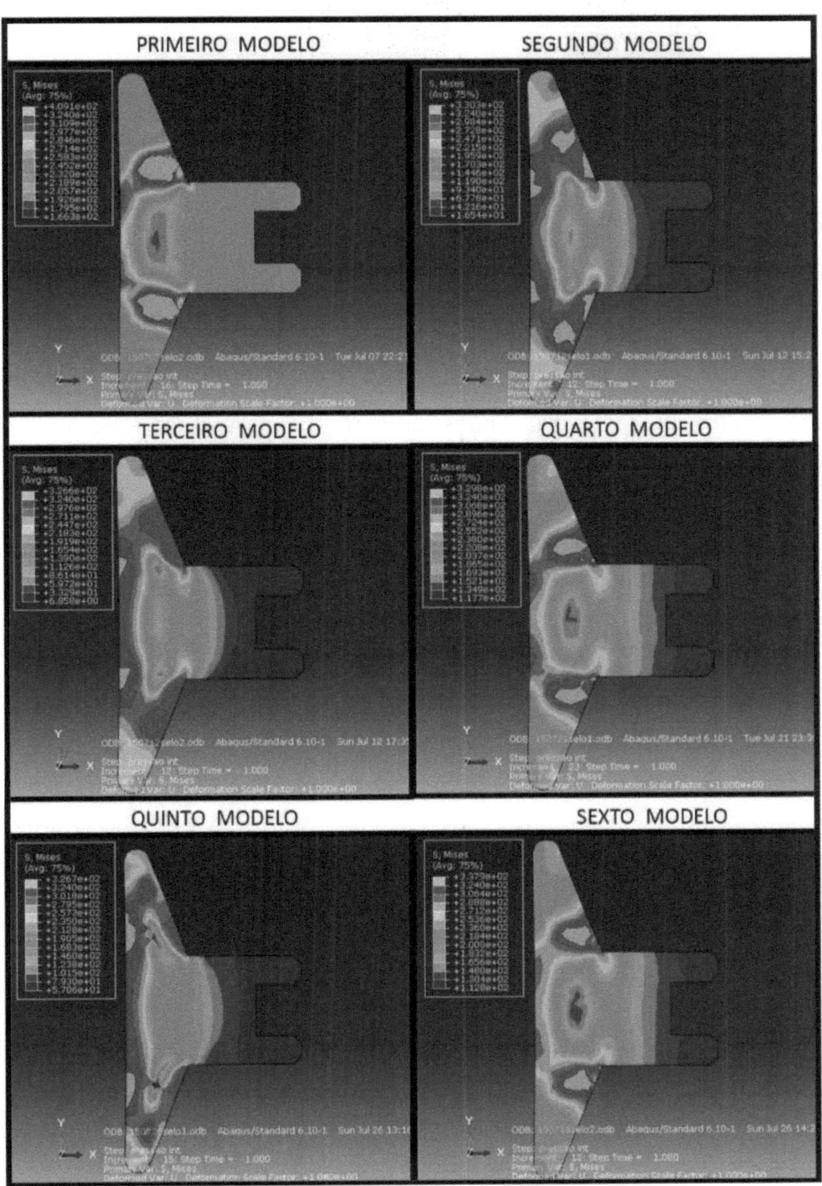

Figure 49 - Von Mises stress in the seal for the six models developed after the final stage [MPa]. The filter was applied to the seal's yield stress value. The regions shown in gray exceed the material's yield strength values.

Figure 49 shows the results for the von Mises stress in the seal, for each

69

model after the final stage. The regions shown in gray exceed the yield strength of the material.

Compared to Figure 40, where the von Mises results are shown after the energization stage in the seal, it can be seen that after the final stage the seals show a reduction in the areas where the yield stress of the material is exceeded.

This is due to the fact that when internal pressure is applied to the model, it tries to separate the upper seat from the lower seat and generates compression on the internal walls. These loads tend to relieve the load on the seal, consequently reducing the stress on it.

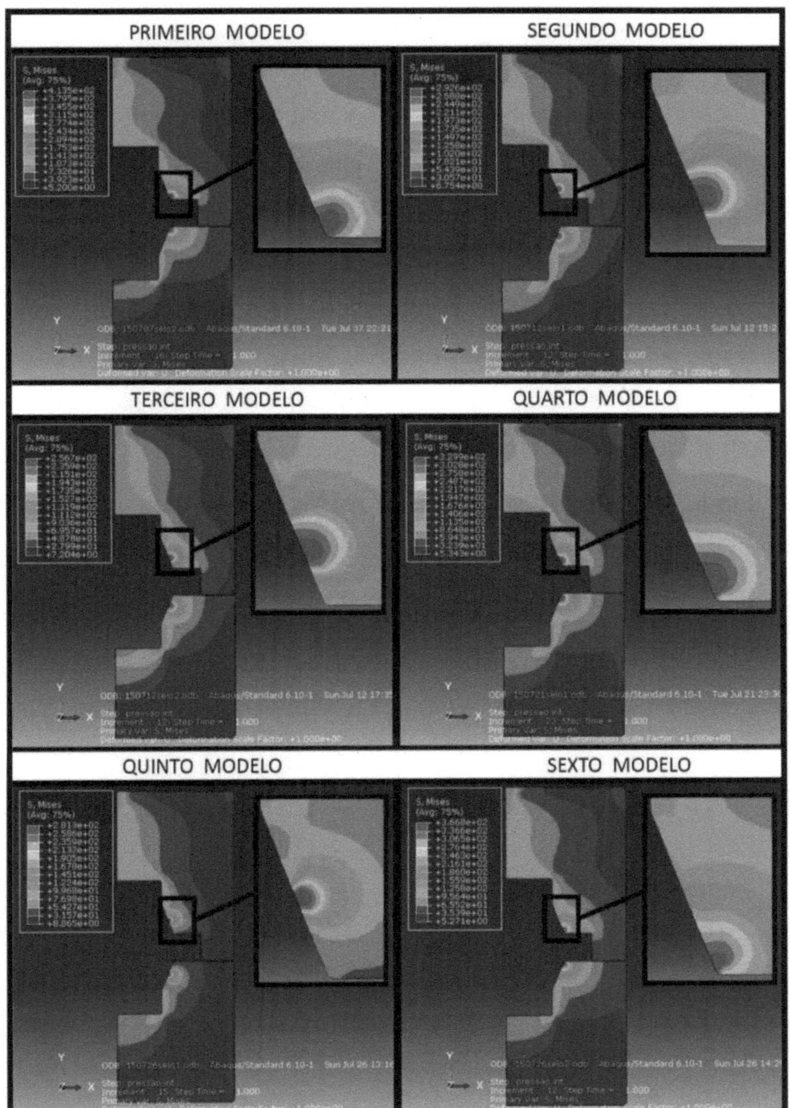

Figure 50 - Von Mises stress in the seats for the six models developed after the final stage [MPa]. The regions shown in red are those close to the effective sealing area, which have the highest von Mises stress values.

Figure 50 shows the results for the von Mises stress in the seats for each model after the final stage. All the models showed an increase in stress in the areas around and in the sealing strip. The first model showed a maximum stress value in the sealing region practically equal to the yield stress value of the material.

Compared to Figure 41, where the von Mises results are shown after the energization stage in the seats, it can be seen that the behaviour near the sealing strips was little affected. Due to the internal pressure applied, the internal areas of the upper and lower seats showed an increase in stress values.

Figure 51 shows the results for plastic deformation in the seal for each model after the final stage. All the models showed an increase in plastic deformation in the areas around and in the sealing strip. The first model had a significantly higher maximum value than the other models.

Compared to Figure 42, where the results for plastic deformation in the seal after the energization stage are shown, there was practically no change. This is due to the fact that permanent deformation occurs in the seal during the energization phase, where the seal is deformed so that it has sufficient contact tension to seal the system.

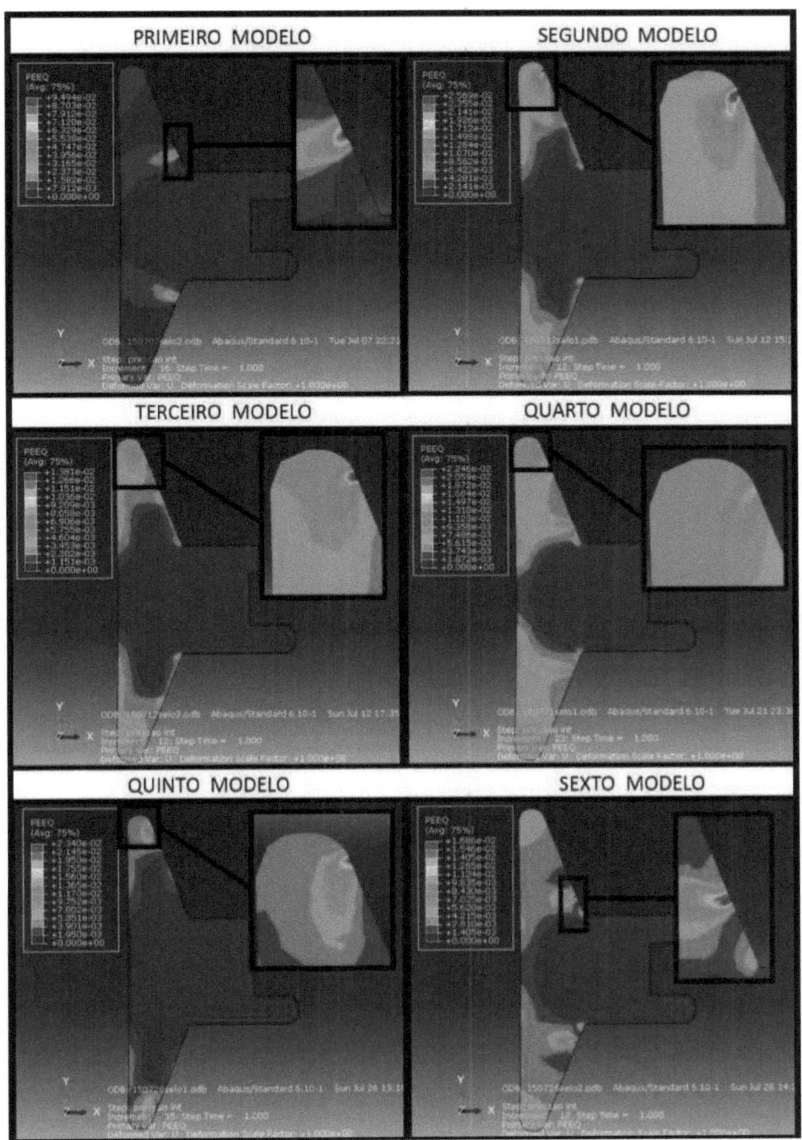

Figure 51 - Plastic deformation in the seal for the six models developed after the final stage [%]. The blue regions showed no plastic deformation. Approaching the effective sealing range of each model, the permanent deformation values tend to show maximum values.

73

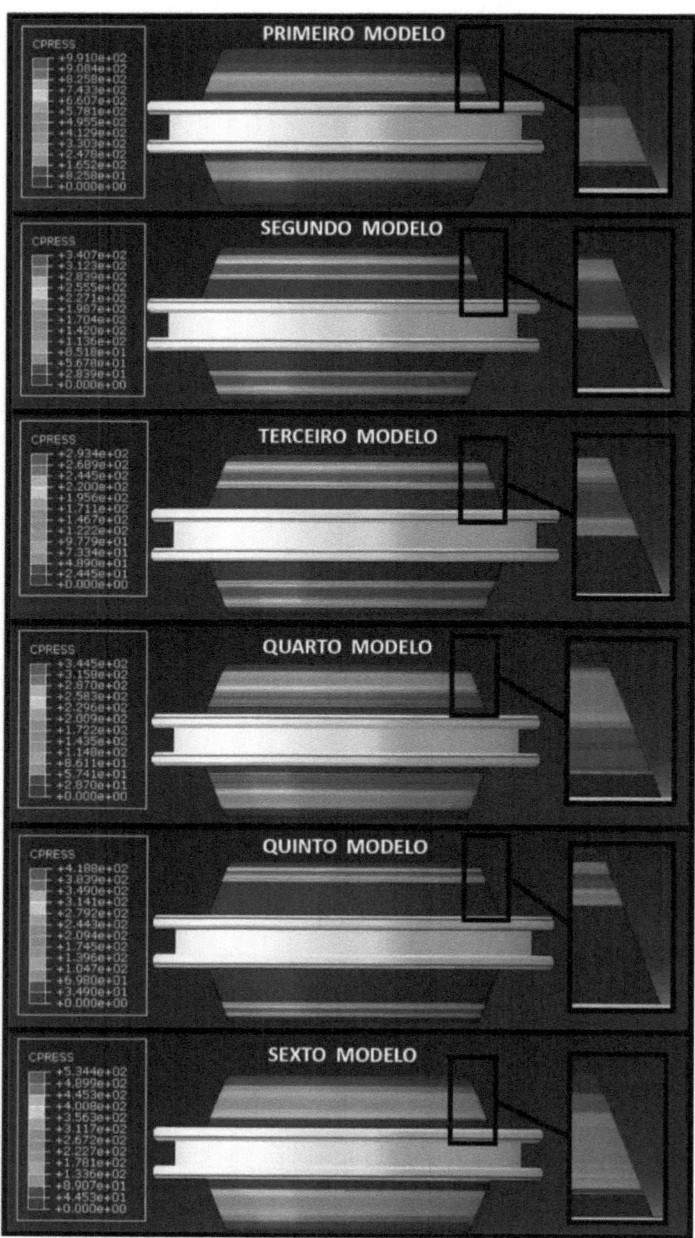

Figure 52 - Contact pressure for the six models developed, after the final stage [MPa]. The regions shown in red are the regions with the maximum contact pressure
values for the seal. These are the regions considered to be the effective sealing range.

74

Figure 52 shows the results for the contact stress for each model at the end of the first stage. The first model had a significantly higher maximum value than the other models.

The region shown in red in each model is the area with the highest contact tension, which is the effective sealing area in the seal: the first, fourth and sixth models have a sealing strip located in the center of the seal; the second and third models have a sealing strip in the middle of the seal, i.e. between the center and the end of the seal; the fifth model has a sealing strip at the end of the seal.

Compared to Figure 43, which shows the results for the contact voltage for each model after the energization stage, the general behavior of the seal and the system in terms of contact pressure remains the same.

4.2. ANALYSIS OF RESULTS

4.2.1. Analysis of von Mises stress results

Another important factor, in addition to the maximum von Mises stress values found in each model, is to visually check which areas of the seal have exceeded the yield strength of the material.

The highest von Mises stress values for the seals occurred in the post-energization stage. Figure 40 shows the von Mises stress for each seal of each model after energization, filtered by the value of the yield strength of the material. All the regions that exceeded the yield strength of the material are shown in gray. Comparing the seal models, it can be seen that the first model exceeded the yield stress almost entirely. The other seals developed tended to have smaller regions that exceeded the yield strength, which were closer to the sealing area.

Evaluating the von Mises stress results for the seats, only the first model approached or exceeded the yield stress of the material. In the post-energization stage, as can be seen in Figure 41, the seat of the first model slightly exceeded the yield stress near the effective sealing area. After the final stage, as can be seen in Figure 50, the stress value was slightly below the material's yield stress.

4.2.2. Analysis of plastic deformation results

Figure 51 shows the plastic deformation for each seal of each model after the final stage.

Comparing the seal models, it can be seen that the first model has larger regions which show permanent plastic deformation and reaches a higher absolute maximum value. The other models show a reduction in the regions showing plastic deformation and a reduction in their absolute maximum value.

4.2.3. Analysis of contact pressure results

Figure 52 shows the contact pressure for each seal of each model after the final stage.

Comparing the seal models, it is possible to see that all the models are capable of sealing, and it is possible to visualize the effective sealing range in each model. The first, fourth and sixth models have an effective sealing range closer to the center of the seal, while the second, third and fifth models have an effective sealing range closer to the end of the seal.

4.2.4. General analysis of the results

By evaluating the results, it is possible to identify the effectiveness of the process when using the finite element method to develop metal x metal seals in an interactive way. After analyzing each model, it is possible to see how the final result was modified after each change made to the model.

Comparing the models, it is possible to identify:

4.2.5. In the first model, the results show a system capable of sealing. The sealing coefficient is the highest among the models. However, the plastic deformation in the seal is high and the seal exceeds the yield stress almost entirely. Furthermore, the first model is the only one with plastic deformation in the seat. Due to the high value of plastic deformation in the seal and plastic deformation in the seat, the use of this first seal is not recommended.

4.2.6. The other models were able to seal. All the other seals showed less plastic deformation when compared to the first model. The other models also showed no plastic deformation in the seats.

4.2.7. Excluding the first model, the two models with the highest sealing coefficients are the fifth and sixth.

4.2.8. Comparing the fifth and sixth models, they have an effective sealing range in different regions. While the fifth model has an effective sealing range at the end of the seal, the sixth model has an effective sealing range near the center of the seal. According to Sweeney et al. (2004), the differential angle between seal and seat is intended to create greater contact tension in the sealing band, this sealing band being located at the end of the seal. The fifth model shows this type of expected result.

4.3. COST ANALYSIS

The purpose of this cost analysis is to present a comparison of man-hour costs between qualification using the finite element method and validation tests and qualification using only validation tests.

Qualification for "metal x metal" seals follows the steps described by API 17 D (American Petroleum Institute, 2011).

In this comparison, only the man-hour costs of each method are taken into account.

Only estimated figures are shown for each method, since each method can vary during execution.

The following steps were considered for the method using finite elements and validation tests:

- Seal design;

- Seal analysis (1);[a]

- Modifications to the seal, according to the results of the analysis (ia);

- Seal analysis (2a);

- Modifications to the seal, according to the results of the analysis (2a);

- Seal analysis (3a);

- Validation test.

The following steps were considered for the method using only the validation tests:

- Seal design;

- Validation test (1st);

- Modifications to the seal, according to the validation tests (ia);

- Validation test (2nd);

- Modifications to the seal, according to the validation tests (2a);

- Validation test (3rd).

Considering the aforementioned scenarios and estimating the number of hours for each activity involving each specific type of professional, it is possible to present a comparison, as shown in Table 13.

Table 13 - Estimated working time for MEF and validation tests and qualification using only validation tests.

Stage	Professional	FEM and validation tests [h]	Validation tests [h]
Stamp design	Design engineer	16	16
Seal analysis (1)[a]	Calculation	40	Not applicable

Stage	Professional	FEM and validation tests [h]	Validation tests [h]
	engineer		
Modifications to the seal, according to the results of the analysis (1)[a]	Design engineer	8	Not applicable
Seal analysis (2)[a]	Calculation engineer	24	Not applicable
Modifications to the seal, according to the results of the analysis (2a)	Design engineer	8	Not applicable
Seal analysis (3)[a]	Calculation engineer	24	Not applicable
Validation tests	Test engineer	80	80
	Technical	80	80
Modifications to the seal, according to	Design engineer	Not applicable	8
the validation tests		Not applicable	80
(1a) Validation tests	Engineering		
	tests	Not applicable	80
(2°)	Technical		
Changes to	Engineer	Not applicable	8
seal, according to the	designer	Not applicable	80
validation tests (2a)	Engineering		

Validation tests	tests	Not applicable	80
(3°)	Technical		
Total per	Engineer	32	32
	designer	88	Not applicable
professional	Engineering		
	calculation	80	240
	Engineering		
	tests	80	240
	Technical		
Total per group of	Engineer	200	272
professionals	Technical	80	240

Table 13 shows the number of hours estimated for each type of professional at each stage of the process. As MEF and validation tests are empirical development methods, these figures may vary.

According to Bay and Bay (2010, p. 180), the cost for an *onshore* test engineer is around US$1,000-1,500 per day. The values considered for each type of professional shown in Table 14 were based on these figures.

Table 14 - Estimated cost per type of professional.

Type of professionals	Daily rate [US$]	Hour [US$]
Engineer	1.250,00	156,25
Technical	625,00	78,13

Table 14 considers the daily rate for an engineer based on the average rate mentioned by Bay *et al.* (2010, p. 180). The value of the technician considered is half that of the engineer. For cost per hour, it is assumed that each professional works 8 hours a day.

Considering the values mentioned in Table 14, the estimated costs based on man-hours are shown in Table 15.

Table 15 - Estimated cost based on man-hours for each method.

MEF and validation tests [US$]	Validation tests [US$]
37.500,00	61.250,00

Table 15 shows that the cost of considering FEM and validation tests is lower than the process that only considers validation tests.

It is important to mention that some costs have not been included in this analysis. In the case of carrying out real tests, the costs of repeating the tests, manufacturing new seals, new devices and the consumables used are not included. When using finite elements, the costs of purchasing a compatible computer, buying or renting FEM *software* and training professionals to use the *software* are not taken into account.

CHAPTER 5 - FINAL CONSIDERATIONS

5.1 CONCLUSIONS

After evaluating the results, all the models showed sealing capacity, since after energization all the seals showed contact pressure higher than the internal pressure applied and after applying internal pressure, they continued to show contact pressure in the sealing region with values above the pressure applied internally. Using the finite element method during the design process, it was possible to improve the seal's performance at the most important points:

• Increased contact pressure, consequently increasing the sealing coefficient;

• Decreasing von Mises stress values. It is desirable to work with the lowest possible von Mises stress values in order to avoid permanent deformation and increase the service life of the sealing system;

• Reduction of permanently deformed areas, consequently increasing the service life of the sealing system.

By improving the seal's performance using the finite element method, the risks of failure during actual validation tests are reduced. Consequently, execution time and costs are reduced.

In conclusion, all the models are capable of sealing. As shown in the results analysis chapter, the fifth model developed is the most suitable for use.

Using the finite element method as a process for verifying and improving seal performance reduces the number of times real tests have to be carried out, consequently reducing costs and development time for "metal x metal" sealing systems.

5.2 . SUGGESTIONS FOR FUTURE WORK

The main suggestion for future work is to carry out validation tests on the calculated seal, comparing the results obtained by the method used in this work with the validation tests. This method of developing a seal using FEM and then carrying out validation tests on elastomeric seals was carried out by (Sinha et al., 2011).

In the design of metal-to-metal seals, one of the steps is to define the dimensional tolerances for the manufacture of the seal and seats. For this reason, a second suggestion is to carry out the analysis described in this paper for two different models, one considering maximum material conditions and a second model considering minimum material conditions.

A third suggestion for future work is to carry out a study on the influence of surface roughness for this type of metal-to-metal seal.

A final suggestion is to carry out the modeling and mathematical formulation of this work, in order to compare the analytical results with the results found by the finite element method presented in this work.

REFERENCES

Alves Filho, A., 2007. Finite Elements: The Basis of CAE Technology, 5th ed. Èrica, Sao Paulo.

American Petroleum Institute, 2011. API Specification 17D: Design and Operation of Subsea Production Systems-Subsea Wellhead and Tree Equipment, 2nd ed. American Petroleum Institute, Washington, DC.

American Petroleum Institute, 2010. API Specification 6A: Specification for Wellhead and Christmas Tree Equipment, 20th ed. American Petroleum Institute, Washington, DC.

Anderson, E., 1996. Nonlinear Finite Element Analysis and Seal Design Optimization. Parker Hannifin Corporation, Salt Lake City, UT.

Augusto, R.A., 2012. Object-Oriented Architecture for High-Order FEM with Applications in Structural Mechanics. State University of Campinas.

Azevedo, A.F.M., 2003. Finite Element Method. University of Porto, Porto.

Bay, Y., Bay, Q., 2010. Subsea Cost Estimation, in: Bay, Y., Bay, Q. (Eds.), Subsea Engineering Handbook. Gulf Professional Publishing, Houston, TX, pp. 159-192.

Beer, F.P., Johnston Jr., E.R., Dewolf, J.T., 2007. Strength of Materials, 4th ed. McGraw-Hill do Brasil, Sao Paulo.

Brandao, M.O., 2007. Contact Analysis of Rough Metal Surfaces Applied to Locking Ball Valves. Federal University of Rio de Janeiro.

Bryant, M.J., 2013. Running-in and Residual Stress: Finite Element Contact Analysis of as Measured Rough Surfaces and Comparison with Experiment. Cardiff University.

Callister Jr., W.D., 2006. Material Science and Engineering, 7th ed. John Wiley & Sons, Inc., New York, NY.

Christensen, R., 2015. How Do Mises and Tresca Fit In. Aeronautics and

Astronautic Dept. Stanford University, Stanford, CA.

Dassault Systemès, 2015. Abaqus® Overview [WWW Document]. URL http://www.3ds.com/products-services/simulia/portfolio/abaqus/overview/ (accessed 3.28.15).

Dril-Quip, 2015. DX® Series Subsea Connectors. Drill-Quip, Inc., Houston, TX.

EagleBurgmann Germany Gmbh & Co KG, 2012. EagleBurgmann supplies high-pressure seals for Russia's ESPO pipeline project. Seal. Technol. 3, 10-12.

Finney, R.H., 2012. Finite Element Analysis, in: Gent, A.N., Ellul, M.D., Finney, R.H., Hamed, G.R., Hertz, D.L., James, F.O., Lake, G.D., Miller, T.S., Campion, R.P. (Eds.), Engineering with Rubber: How to Design Rubber Components. Carls Hanser Verlag, Munich, pp. 257-305.

Flach, P.M., 1995. A seal is born. World Pumps 351, 48-50.

Hou, M., Su, M., Liu, Y., Gui, Z., 2010. Analysis of a dovetail O-ring groove performance. Seal. Technol. 8, 9-12.

Johnson, K.L., 1987. Contact Mechanics. Cambridge University Press, Cambridge, England, UK.

Jong, I.C., Springer, W., 2009. Teaching von Mises Stress: From Principal Axis to Non-principal Axis. American Society for Engineering Education, Washington, DC.

Lima, L., 2015. Strength Criteria, in: Lima, L. (Ed.), Strength of Materials IV. Faculty of Engineering, Rio de Janeiro State University, Rio de Janeiro.

Manzoli, P.R.P., Pastoukhov, V., Ramos Jr., I.D.C., 2014. Simulation of Crack Propagation in a New Antenna Installation on the Fuselage of a Commercial Aircraft, in: XI Symposium on Computational Mechanics. Juiz de Fora, MG.

Natal Jorge, R.M., Dinis, L.M.J.S., 2005. Plasticity Theory. Faculty of

Engineering, University of Porto, Porto.

Sinha, N.K., Raj, B., Mukhopadhyay, R., 2011. Development of fluoroelastomer back-up seal: snapshots of a novel effort in rubber engineering for a critical application. Seal. Technol. 5, 9-12.

Special Metals Corporation, 2006. Inconel® Alloy 625 [WWW Document]. URL

http://www.pccforgedproducts.com/web/user_content/files/wyman/Inconel alloy 625.pdf (accessed 3.28.15).

Special Metals Corporation, 2004. Incoloy® Alloy 825 [WWW Document]. URL

http://www.pccforgedproducts.com/web/user_content/files/wyman/Incoloy alloy 825.pdf (accessed 3.28.15).

Sweeney, T.F., Brammer, N., Chalmers, G., 2004. Gasket with Multiple Sealing Surfaces. US 6772426 B2.

Trelleborg Sealing Solutions (TSS), 2013. Sealing successful FEA. In the Groove 28, 6-8.

Vartziotis, D., Wipper, J., Papadrakakis, M., 2013. Improving mesh quality and finite element solution accuracy by GETMe smoothing in solving the Poisson equation. Finite Elem. Anal. Des. 66, 36-52.

Zhao, C., 2011. Ultrasonic Motors: Technologies and Applications. Springer, Beijing.